WILD WINTER

ALSO BY JOHN D. BURNS

The Last Hillwalker

Bothy Tales

Sky Dance

WILD WINTER

In search of nature in Scotland's mountain landscape

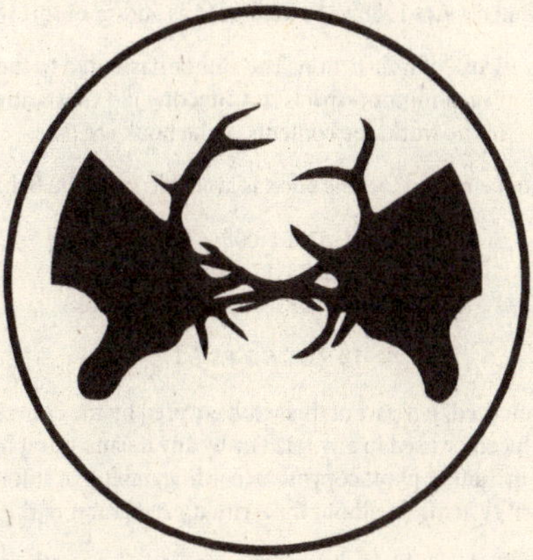

JOHN D. BURNS

Vertebrate Publishing, Sheffield
www.v-publishing.co.uk

WILD WINTER

JOHN D. BURNS

First published in 2021 by Vertebrate Publishing.

VERTEBRATE PUBLISHING
Omega Court, 352 Cemetery Road, Sheffield S11 8FT, United Kingdom.
www.v-publishing.co.uk

Cover and title page illustrations by Jane Beagley.

A CIP catalogue record for this book is available from the British Library.

ISBN: 978-1-83981-005-3 (Paperback)
ISBN: 978-1-83981-006-0 (Ebook)
ISBN: 978-1-83981-007-7 (Audiobook)

10 9 8 7 6 5 4 3 2 1

Cover design by Jane Beagley, Vertebrate Publishing.
Production by Cameron Bonser, Vertebrate Publishing.
www.v-publishing.co.uk

Vertebrate Publishing is committed to printing on paper from sustainable sources.

FSC www.fsc.org MIX Paper from responsible sources FSC® C018072

Printed and bound in Great Britain by Clays Ltd, Elcograf S.p.A.

CONTENTS

~

To see a World in a grain of sand,
And a Heaven in a wild flower,
Hold Infinity in the palm of your hand,
And Eternity in an hour.

When my father died in his ninetieth year, I found these words in his photograph album amongst his images of wild flowers and walks in Wales, written in his elegant handwriting. William Blake's lines convey the love of nature that my father and I shared, and I hope that this small book captures some of that sense of wonder.

~

OCTOBER

The darkness consumes everything. As I walk up the hill, my breath mists in the beam of my head torch and the silence deepens. There is no birdsong and no wind to disturb the stillness of the night. The world is reduced to silhouette. The trees of the approaching woodland appear only in outline, and in the distance mountain ridges lie black against the vastness of the star-speckled sky.

At last I reach the forest and can bury myself amongst the rough tree trunks. When I turn off my head torch, the darkness becomes tactile and the silence grows and pools about me. Robbed of sight and sound, my brain craves sensation and I lean my face against the bark of a pine tree. There is reassurance in the smell and feel of the coarse bark against my cheek. As I sit amongst the soft, damp moss, I become acutely aware of my breath, which slows and becomes shallower now that I have stopped climbing.

Sitting in the dark, I am acutely aware that the slightest sound will carry far in this still air. I must be as silent as possible to find my quarry. This October night is cold. The summits of the hills are glazed with frost. The cold air sinks down from the heights and chills me as it fills the glen. With vision lost, my ears strain to catch the tiniest sound that might reveal a movement out in the blackness. My imagination creates illusory sounds. Was that a twig breaking? Did the grass rustle just then? Perhaps it was just the night moving.

Below, the Highland glen of Strathconon weaves its way into the horizon. To the east, the mouth of the glen opens out towards the market town of Dingwall. To the west, the fingers of the glen reach out to the outlying hills. Beyond the ridge lies Achnasheen, from where this chain of mountains rolls out to the sea loch at Lochcarron, and further still into the wild Atlantic. I know this place well. In my imagination I take an eagle's ride over the sweeping ridges, across the dark lochs and down the wide glen to where the lights of houses twinkle at the roadside. The journey is filled with memories of days spent wandering in the rain, days on sunlit rock climbs, days on snow-crusted hills – some with friends and others alone with the landscape. These valleys and hills keep drawing me back. Here I am once more in this familiar glen, waiting for another day.

A glow begins to form in the V of the mouth of Strathconon. The light of the new day is creeping across the horizon and the features of the glen are slowly emerging. In the pale dawn, I struggle to make sense of light and shadow as colour gradually seeps out of the darkness, like a photograph developing. Soon I see a rocky crag above me; below, the dark wound of a stream bed winds across the floor of the shallow corrie. The shape of the landscape is no longer obscured as the reluctant night leaves the valley.

I miss too many dawns. I spend them idly in bed, or waste them buttering toast and bumbling about the internet. I am too concerned with the gibbering press or reading my mail, bleary-eyed. I never notice that outside a miracle is occurring. A new day is slowly coalescing into life. Every time I see a dawn, I vow that I will watch more of them, and yet somehow life distracts me and I forget to make space to wonder. At least I am here for this dawn. Now it will not tiptoe past unseen. The darkness is yielding, letting the burning colours of autumn slip through its fingers. Still I wait, listening to the forest breathing in the early morning breeze. I have been sitting amongst the trees for over an hour and it is difficult to keep warm. I shiver as icy fingers find their way inside my jacket. I try to ignore the cold, forcing myself to remain motionless, knowing that the slightest movement could mean that my nocturnal vigil has been wasted.

At last it comes, drifting through the trees: a deep, guttural, primeval roar. A grunting yell that has echoed through these trees and across these hills since the ice retreated thousands of years ago. A shape moves in the semi-dark, only to shift back into blackness as my brain tries to make sense of the gloom. Again the roar sounds across the glen, closer this time. After the echoes die, silence returns and the valley falls into a soft stillness. Minutes pass, until I think I may have given away my presence. The roar comes again, but still I see nothing. This time, the sound is followed by the rasping of great lungs filling with air. I can hear hooves picking their way through the boggy grass.

Out of the gloom swaggers a powerful creature, the master of this glen. He is so close that I *feel* the sound of his call vibrating the air as much as I hear it. He shakes his antlers, his breath clouding in the morning air. Seconds later, his challenge is answered by another male anxious to stake his claim. The stag turns his great antlered head and trots away towards his challenger. Now stags come from all directions, bellowing and roaring, each staking his claim on the rutting ground. As the morning grows brighter, the shape of the landscape reveals itself. Below me, the ground slopes away to a small ravine; beyond that, closed in by the hills above, there is a level area the size of half a dozen football pitches. This is where the drama I have come to see will unfold, where the battle of the rut is to take place. Now more and more stags are coming into view. Some beasts are huge and powerful, twice the weight of a man or more. Others are less impressive and will take another year to reach their full strength. These younger animals have no chance of winning the competition to mate with the females. Though they cannot win, the surge of hormones released as the autumn rut arrives compels them to be there.

Soon the amphitheatre echoes with challenges and answering calls from over thirty stags. Many are alone, but others have miniature harems of four or five hinds. The stags that have managed to collect a bevy of admirers have to fight off constant challenges from other males. The rut is an exhausting process for them, and by the end of the season, lasting from the end of September to the third week of October, many stags will have lost a fifth of their body weight.

Deer have acutely sensitive hearing and will flee if I so much as zip up a jacket. From the years of hunting by early man, they have learnt to detect humans by their outline. By staying close to the trees, I disguise the outline of my shape so they cannot see me. The stag that was close to me is challenging an older stag with four or five hinds in tow. He is roaring and shaking his antlers to show how big and strong he is. The old stag turns and faces him, responding to the challenge; if anything, he is larger than the animal I first saw. The pair walk parallel to each other for about ten minutes, each hoping that this display of bravado will intimidate the other into backing down and avoid a fight which could injure or even kill one of them. Although fights are common, stags try to avoid physical confrontations. Such battles are dangerous. A stag that gets an antler in his eye or whose internal organs are pierced will have a long, lingering death. Both the animals I am watching refuse to give way. On some secret cue, they turn and lock antlers. The sound of clashing bone echoes across the hillside. The stags grunt and snort with the effort of combat, their hooves sending grass and sods of earth into the air as they both struggle to keep their footing. They wrestle for almost ten minutes. At first the larger stag pushes the challenger back, his extra weight giving him the advantage. Then he stumbles on the broken ground and almost falls. The smaller animal chooses his moment and hurls himself forward, driving his antlers into his opponent's side. The larger beast staggers back, blood dripping. He tries to mount another attack but his strength has left him, and he turns slowly and walks away. Perhaps his age was against him. It may be that he will never again hold sway over his own family of hinds. Defeat on the rutting ground has left him with a solitary life; he has lost his harem. The smaller stag strolls casually across to the hinds to make their acquaintance.

There are fights and roaring contests breaking out all over the rutting ground as males battle for dominance. Oddly, those not involved in the duels graze peacefully as if all this mayhem has nothing to do with them and they are just enjoying a light breakfast. This spectacle is being repeated all across the Highlands, in remote corries and glens, as it has been for thousands of years. I am watching an ancient ritual.

At last the sun rises, filling the glen with light and revealing the autumnal colours of the trees around me, a patchwork of browns and golds. Beyond the forest lie the green meadows. White-painted houses dot the single-track road that runs to a dead end some seventeen miles along Strathconon. At this point, midway along the valley, I am only some twenty-five miles from my home in Inverness, yet this feels like a different world.

As the day begins, the rut slowly calms and the stags gradually leave the arena and climb away up the hill. This is late in the rut; it will last only a few days longer. The stags I am watching are mainly the smaller, less dominant males. They are fighting to hold sway over the few females who remain after the larger, more dominant stags have finished their rut. In only a few minutes the rutting ground is empty. I climb, stiff with cold, out of my hideaway in the woods. The trees I hid in are Scots pine, once part of the Great Forest of Caledon that covered the Highlands. Although it is long gone, its remnants cling to small pockets, a memory of what used to be.

At this time of year in the Highland hills, the sounds of the rut are a constant companion. The Gaelic people called October 'the roaring month'. I've heard stags bellowing their challenges when I have been sitting by the firesides of remote bothies and on autumnal walks. Yet, until now, I have never witnessed them close at hand, living out the battles that decide the course of their lives. A few minutes after witnessing the battle, I am walking back to the road, crossing the rutting ground. In this small area there are half a dozen tall metal feeding troughs. Seeing so many stags together in one place is unnatural. The numbers in this glen, and many others across the Highlands, are artificially high. Red deer are creatures of the forest, yet here they are on open ground. The hills around me are criss-crossed with tracks, like deer motorways, but this glen would be unable to naturally sustain such large numbers of deer. Calling what I have just seen natural is to misconstrue what nature would have made of the Highlands. In truth, it is more the outcome of human intervention.

The rutting ground is grazed bare, and there are no trees or bushes in this open space. Scotland's landscape has paid a heavy price for rich men's obsession with shooting deer. Right across the Highlands, the land is a patchwork of sporting estates where excessive deer numbers prevent the regeneration of forestry. What happens in our upland areas is based on the whim of a few landowners. As I walk past the metal feeders that the estate fills with hay in the wintertime, I wonder how long this archaic system will last. The feed stops the deer from starving when the weather is cold and food is scarce, but it keeps the deer population unnaturally high. No one can deny how beautiful the deer are. The sight of the stags roaring their challenges across the wild glens is a true spectacle, but the large herds strip the hillsides bare.

Despite their awesome appearance and the ferocity of their calls, even the powerful stags are no risk to people. While they can grow tame, these creatures never lose their instinctive fear of us. I only know of one instance of a stag attacking a human and that must be over thirty years ago now. The tale goes that a walker in the Cairngorms was once picked up not by a red deer but by a reindeer. The animal carried him several hundred metres and only left him alone when the man played dead. I can only think that the reindeer had experienced a moment of transient lunacy.

I am pleased that I witnessed the rut. It is something I will never forget. We hurry past things in our lives far too fast. Taking the time to see the things that we know are there but forget to value is so important.

October is the herald of winter in the Highlands, and has always been a special time of year for me. The month comes in bringing a rash of glorious colour to the glen, along with gales that rampage across the hills, carrying away the last of the summer's memories. The hills themselves turn from green to shades of ochre almost as if they are coming into bloom. October brings an end to the cursed midge, the bane of the hillwalker. The calls of the rut are the herald's fanfare telling all who travel in these high lands that the easy days of warmth and plenty are coming to an end and that they must prepare for the hardships ahead. October sees the first of the winter snows on the high tops. Tantalising brushes of white that often appear first on the high plateau of the Cairngorms or the summit of Ben Nevis linger for a few days and are then washed away in the next downpour of rain. When my passion was for climbing in winter, I could never resist scouring the distant summit of Ben Wyvis, hoping to catch the first dustings of snow that signalled the start of the winter season. Today, once October comes, I cast my eyes every morning towards that long ridge, searching for the merest suggestion of whiteness, just a hint that in the high places winter has already arrived.

***

A week after watching the rut, I am walking along the road that runs through the Ardtornish estate and ends at the small west-coast village of Lochaline. This quiet rolling track, no more than a car's width, was once an important route as it was the only way through the village to the ferry that serves the Isle of Mull. Carts would have rattled along it carrying passengers with long journeys to complete in the Western Isles. Two hundred years ago, some would have been heavily laden with charcoal, heading for Oban to make the cannon that would arm Nelson's fleet. Even today, if you wander through the oak woods that border the loch shores here, you will come across rings that are the remains of the charcoal burners' pits. All the folk of the townships, small clusters of homes scattered in the hills, would have met and gossiped here. Now this road has been forgotten. There is a stillness about this narrow corridor as it passes along the shore of Loch Aline, crowded by the thick woods to the north and bordered by the lapping waters of the sea loch to the south. Bushes and trees press in from either side as nature struggles to reclaim the old road.

I zip up my jacket against the cold sea wind as it chills my neck. I look out across the loch to the far shore, a mile or so distant. There the green of the woods drifts down the low hills, dappled with patches of russet brown where autumn has touched them. I walk quietly, slowly, watching. I am looking for dark shapes in the water, for a head bobbing up or a tail splash. I hope to see otters along this shoreline, perhaps fishing for crabs or devouring sand eels.

Otters are not uncommon in the River Ness, yet they mostly pass unseen through the city. Sometimes I see them in the morning as they hunt, mere metres from the schoolchildren walking to class, the shop workers and the people trudging into the office. Few folk raise their eyes from the pavement or from the Twitter feed that tugs insistently for their attention. No one turns to see the otter's sudden ripple on the water or to catch sight of a tail as it dives for fish. Two worlds sit beside each other, the wild otter's and the tame commuter's, rarely touching. The otter sightings I get near my home are usually only fleeting. Here, on the shore of Loch Aline, I am hoping to be able to get a more intimate understanding of their behaviour. I want to be able to watch them feeding or playing, to know them a little better.

The sight of an otter is always exciting. It is a glimpse of a wild creature that gives me hope that our world is not completely tame. Seeing an otter is to look through a window into a wild world just beyond our reach, beneath the rippling water.

As with all wildlife, finding a chance to watch them in detail won't be easy. The glimpses I have caught so far have been accidental, opportunistic moments when our paths have crossed. Now I must set out to find them, deliberately choose a spot and wait for them to arrive. That is a far greater challenge. Today, as I scan the shoreline with my binoculars, the water seems eerily empty. The sky is dappled with growing grey clouds and the wind picks up and spots me with rain. I walk for an hour or so beside the loch and come across a patch of odd, dense vegetation. Tall reed-like bunches of grass are growing beside the track with green fronds sprouting from their joints. I've no idea what this plant is, but it looks primitive, almost prehistoric. I photograph it with my phone and make a note to look it up when I return to the bothy.

The extent of my ignorance about the environment is embarrassing and frustrating. This winter, I want to learn about the creatures around me. I can only identify a handful of birds and my knowledge of plant life is execrable. I will never be a naturalist, never be one of those people who can pick up a leaf and say, 'Oh, that's *saxifriogis largebergensis*.' I just don't have the kind of mind that can keep lists tucked away in mental drawers and understand how one plant is related to another. I can't hold all that information and produce it out of thin air like a cognitive conjuring trick. That might be my failing – or, of course, the other explanation could be that I'm just lazy. The fact that I'm largely ignorant never used to bother me, but it does now. What's changed is that I have begun to understand that the bond I have with the natural world is important to me. I feel at ease here beside this sea loch, yet I don't really know it or understand its ecology. I need to deepen my knowledge, and I know I can only do that by studying it on cold, windy days like this one and trying to understand what I am seeing.

The wind swirls in from across the dappled water of the sea loch and suddenly the air fills with tiny whirling helicopters. Little seedlings spin down on me like aliens scattering from space, exiting the mother ship to set up colonies on distant planets. I look up through the branches of the tree above me. So this must be a sycamore tree, or maybe it's a beech tree. I'll look it up when I get back to the bothy too.

After a couple of hours with no sign of otters, I decide that it is not the day for them here and wander back along the old road towards Ardtornish. I might stand a better chance of seeing otters by driving along the narrow coastal road that runs to Drimnin. There the coast is a lacework of pools and islands, offering them ample hunting grounds.

As I turn to leave the lochside, there is a commotion in the air above me as two gulls pursue a buzzard. The buzzard sweeps and dives to escape, but the gulls are equal to its gymnastic flight and it struggles to elude them. The gulls press home their attack, sweeping in one after another to strike. The buzzard fights back, turning towards the gulls with his sharp talons extended. But the gulls are a determined foe and eventually the raptor is forced to seek refuge amongst the branches of the oak wood. The gulls only relent when they are sure that the bigger hunter has been banished from their territory. I feel privileged. There must be many thousands of folk in offices and factories all across the country who would have been glad to have seen this.

The Drimnin road leaves the village of Lochaline and winds its way along the westernmost edge of the peninsula. On my right are woods so dense they seem impenetrable, home to that most elusive of creatures, the wildcat. To my left, open fields lead down to the sea. A few miles away sits the wild Isle of Mull, its rugged hills rising dark and green out of the water. The wind is sweeping great billowing clouds, as big as the hills themselves, in from the west. They tumble over the mountains of Mull to crash into the sea beyond. The weak sun fights its way through clouds and brings a surreal glow to these great mountains of mist.

The Drimnin road is always quiet. It extends for about ten miles from Lochaline and serves only the few white-painted houses scattered along its length. There is barely a car every half hour, and sometimes not even that. I squeeze my car off the road at a place known to the locals as Dog Island. It's low tide right now and I can walk to the island, which is about the size of two football pitches and only a few metres offshore. Stepping out of the car, I am surprised by the chill wind. There is ice on the small pools as I make my way across. Normally I'd be sinking inches into the bog, but now the ground is frozen and I can stand on it. An old stone wall leads past a dilapidated boathouse with sheets of corroded corrugated iron hanging from it like the tattered clothes of a scarecrow. The island is a patchwork of pools and small bays, perfect otter habitat, but each bay is empty. The remains of their meals are everywhere, the broken pink shells of crabs scattered about as if left by careless partygoers, but the diners are absent. It seems today is not going to be my day for seeing otters. I head back to the bothy as the light begins to fade.

On the way back, I call into the village to pick up some supplies at the village shop. Lochaline is typical of communities on the west coast of Scotland. There's a school, a pub, a church and even a doctor's surgery. Most important of all, there's a shop, a low white building that always confuses me as the front doors are frequently locked and the side entrance looks as though it should be closed but isn't. As I walk into the store, a small group of locals are gathered round the counter. They stop talking the instant I walk in. I'm at that awkward stage in a small place like this. I'm not quite a stranger because they've all seen me before on my frequent stops here. If I was a stranger, I'd be an irrelevance since I'd be just passing through and would have no idea who the people were talking about. The problem is that I look familiar but most of them can't quite place who I am. That makes everyone, including me, feel awkward. Half a dozen pairs of eyes are watching me and trying to decide if I'm that newly retired man from the cottage on the shore or if I'm working on one of the houses being built nearby.

Mary, the middle-aged woman behind the counter, beams at me. She knows who I am because I told her on a previous visit so that she'd be able to tell everyone else and save a lot of trouble.

'Is that you away up to the bothy?' she calls over to me.

This is not a question. Mary isn't even talking to me. She is telling everyone else in the shop who I am. Now they know I am 'one of those who go up to the bothy' and everyone will know how to treat me. They don't have to ignore me, because I'm more than just a visitor, but I'm not local, so they can speak in front of me. They can talk to me about weather and all manner of trivia, but there's no point in talking to me about the drinking habits of the local minister or the latest unexplained pregnancy. I'm no use for that type of important local gossip.

I smile as I pick up firelighters and soup. 'Aye, that's me.'

This general store has one version of everything from wellies to baked beans. There's no choice about what brand you buy because there simply isn't room in the store, so if you want beans it's those beans you have to have. As I leave, the little group at the counter goes back to fulfilling the main role of the shop, which is the exchange of gossip. Some people come to isolated places like Lochaline to hide themselves away from the world.

The truth is that there is far more anonymity in any city than there ever is in a west-coast village.

As I leave the shop, laden with firelighters, tinned soup and bread, I smile to myself because I can see people here that no one else can see. I'm not hallucinating. This is where I set one of the scenes of my novel *Sky Dance*. I can see Tabatha Purdey in the shop doorway, barging into Angus and Rory as they head off to their bothy. Tabatha, weeks prior to this meeting, had discovered the two hillwalkers naked beside a river having narrowly escaped drowning. When she meets them in the shop, she knows she has seen them before but can't quite place them. Angus and Rory are mortally embarrassed, hoping that Tabatha won't remember. The three characters are as real to me as the other folk in the shop.

Twenty minutes later, I'm standing in the car park at the foot of the hill, packing my rucksack for the walk in to the bothy, when a van pulls up beside me. It's Dave who works part time for the Scottish Wildlife Trust and helps to manage the nature reserve here. He steps out of the van and is instantly followed by his sheepdog, Jess. She remembers me from my volunteering on the nature reserve and rushes over to bury her muzzle in my hand with the kind of joyous friendship only dogs seem to be able to muster.

Dave spends most of his working life outdoors and has a quietness about him that echoes the peace he finds there. He's a slim, fit-looking man with light hair. We chat for a few minutes about the weather and this and that, but mostly about how cold it is. The ever-changing Highland weather always gives us plenty to talk about.

Dave tells me he's come to walk his dog. 'I'll walk with you as far as the bridge if you like.'

The bridge is a couple of miles up the glen about a mile short of the bothy, and I am glad of Dave's company as I have been alone for most of the day. The path climbs steeply at first for a few hundred metres before it levels off beside the river gorge. The hills of Ardtornish are treeless, as are most Highland estates apart from the odd forestry planation. Here the sheer-sided gorge has prevented grazing animals from nibbling off new growth, and its sides host dense trees and bushes. I've often seen the tracks of pine martens on my way to the bothy. I'm sure they live in the thick cover of this gorge and I'm keen to see one, although I've not been lucky so far.

Dave is one of those people who put my knowledge to shame. 'Look high up – at the edge of the woods, you can see new growth.'

I see young trees on the edges of the older woodlands high on the slopes. 'Yes, I see it. I never noticed that before.'

'The estate have reduced deer numbers. The place is beginning to regenerate,' Dave tells me.

The ranger's eyes are educated; mine are not. I'm enjoying the chance to chat and perhaps pick up the odd bit of knowledge. As we walk higher up the path, we are sheltered from the wind by the oak trees. Jess revels in the walk, running ahead and then darting back to us. She covers the ground effortlessly.

This early in October, most of the trees are still green. A few scattered acorns that crunch beneath my boots give the game away that the year will soon be ending.

'I've not seen anywhere with so many oak trees.' I'm curious as to why oaks live here when I've rarely seen them in the rest of the Highlands.

Dave pats the trunk of one of the oaks with real affection as if he were patting his dog. 'We are in a maritime oak wood,' he explains. 'This is rainforest.'

'I thought rainforests only existed in tropical places like the Amazon.'

Dave laughs. 'Aye, well, we've got a wee bit of our own rainforest right here. These oak woodlands here and those that run around the coastal areas of Ardornish trap the moisture so they're always wet.'

'I never thought of it that way.'

I think back over all the times I've been to Ardtornish and walked through this woodland. Rain is a constant companion here. The Ardnamurchan peninsula is not the place to go if you are worried about getting wet. Even when it is not raining, the woodland air bursts with moisture. The mosses sparkle with water droplets and the trees constantly drip water on to you. The mild climate and the abundance of water lead to luxuriant growth in the woodlands. The trees are sometimes so thickly covered by moss that their bark is hardly visible. The branches of the trees hang with lichen feeding itself from the air.

It's late afternoon by the time we reach the wooden bridge across the river where the gorge ends and the rest of the track continues over open country.

We are standing for a moment, leaning on the wooden rail of the bridge and looking down into the churning waters of the river, when Dave suddenly looks up and points. 'Look there.'

A white wing flashes out in the long grass beyond the river.

It vanishes and I am wondering if I imagined it when the bird rises a few feet above the grass and sweeps towards the bridge, following the rise and fall of the ground with incredible grace.

Dave has the bird in his binocular sights in an instant. 'Male hen harrier,' he says. I can hear the excitement in his voice.

The bird is light underneath, a little larger than a buzzard, with dark tips to its wings. I fumble with my binoculars, lacking Dave's ability to produce them as if from thin air. Once I see the harrier, I too marvel at the graceful creature. There is something ethereal about the way the bird moves. It is part of the air. In seconds it has vanished down the gorge.

Dave is beaming. 'I've not seen a harrier here for two years.'

'Are they rare?'

'Well, yes. But they shouldn't be. They are persecuted on grouse moors. The white ghost.'

We chat for a while and then I shoulder my rucksack and Dave heads back down to the car park with Jess skipping about his heels. As I walk the last mile or so to the bothy, I reflect on how excited I am at the sight of the harrier. A glimpse of wildness – what a privilege today has been.

***

The bothy is a simple place. It sits on the far side of the river across an increasingly rickety bridge which will soon succumb to the rising waters of a winter storm. There is a stove that burns coal to heat the room. For lighting, I have paraffin lamps and candles. To give an air of luxury, there are two armchairs. I sit down, set up my phone and folding micro keyboard, and begin to make notes on the past day. There are a few things I ought to tell you about bothy writing before you pack up your word processor and head for the door. It is not the rural idyll it sounds. Before I can begin writing, I have to put on my writing clothes. These are a heavy pair of fleece trousers, a serious Everest-type duvet jacket and a woolly hat. Despite all these precautions, it is not long before the cold begins to nibble at my extremities. I may shortly have to resort to sitting draped in my sleeping bag wearing fingerless gloves, like some new-age Dickensian miser. This October, it's about 8° Celsius in the bothy, the same temperature as it was in December last year, when I also exiled myself here to write. I have a small supply of coal, but I am conscious that it needs to last most of the winter, so I resist the luxury of lighting the fire, putting off that moment of joy until around 5 p.m. each day.

The problem is that writing is by nature a sedentary pursuit. Eight degrees is not that cold, but writing forces me to sit still for long periods, and that inactivity means that the cold inevitably seeps in. The weather has been strange this year in the Highlands. Normally we get two sorts of winter: a mild, wet winter, when it rains endlessly and gales drive us from the hill; or a cold, dry winter, when the hills are deep in driven snow and frost grips the glens at night. Last winter was different from all the others I have experienced here. It was a mild, dry winter, so dry that many of the hydroelectric schemes across the Highlands struggled to produce sufficient power. Early January brought a week of sub-zero temperatures when the hills were covered with snow and it felt as if winter had at last arrived. The cold did not last, however, and by February the snows had gone and we were left with a black winter, when the hills never turn white.

The mild winter gave way to a cool, wet summer when most of the year's rainfall fell. Such aberrations in climate are death to wildlife. In the previous summer, much more typical, with at least its fair share of hot, sunny days, the glen was a very different place. Last year, warm summer meant the air was buzzing with insects and the valley alive with birdlife. Birds of prey, eagles, hen harriers and merlins were frequent visitors then, and the house martins nesting beneath the eaves of the bothy produced two broods of youngsters. The adults were endlessly on the wing, sweeping out from the little mud nest to return again and again with their beaks bulging with insects to feed their ever hungry and demanding offspring. For several months, we were cursed by hordes of horseflies, vicious biting insects capable of drawing blood even through clothing. In those months, if you stood still, these annoying creatures would zero in on you and drive you from the spot with painful bites that itched for days. Once, unable to bear their ministrations any longer, I bent to take some insect repellent out of my rucksack. This was what they had been waiting for and four of the cursed creatures bit me in the backside simultaneously. I ran screaming to the bothy which provided the only respite as, like midges, horseflies will not enter a bothy in any great numbers, a fact for which I, along with multitudes of bothy-dwellers, am infinitely grateful.

Sitting in the bothy now, my breath misting in the air and my hat pulled round my ears, I think back to the summer we've just had. I don't think I was bitten by one horsefly and even the pestilential midges were subdued. Whilst the insect hordes may be a curse to the rambler, they are a crucial source of food to much of the glen's wildlife. For the house martins, the lack of insects and the cold conditions meant it was challenging for them to raise even one brood of chicks, let alone a second in late summer, and by the end of July the mud nests beneath the bothy eaves were empty and silent. The weather and the paucity of food meant that many fledgling birds struggled to survive.

For the raptors, too, times were hard. The vole population, small mouse-like creatures who are normally here in abundance, was also hit by the lack of insect life. Their part in the food chain is to feed the owls and the hen harriers, who were also notable by their absence in the glen. This makes our sighting today even more memorable. The year that is coming to a close has been a difficult one for wildlife.

As I begin to write about my day, images pass through my mind. I see the waters of Lochaline and feel the wind chilling my cheek. I recall the remains of the crabs after the otters had eaten their lunch, and the vast sky full of towering clouds rolling across the hills of Mull. The one image I see more than any other is the magical sighting of the hen harrier dancing over the grass, weaving its mystical tale in the wind. It is a sight that will stay with me forever.

I wonder if I can tell the story of a winter, my winter here in the Highlands. The days out in the rain. The cold, the adventure and the frustration of not finding the creatures I seek. In this book, I will tell the story of one winter, the coming season. I do not know what sort of winter lies ahead. With the damage done to the environment, it is becoming impossible to predict. The tabloids have forecast the coldest winter for a hundred years, but then of course they always do. I hope to be able to record something of the trials of this season for both people and animals. Let's see what adventures I can find.

When I am writing in the bothy, my life follows a simple rhythm. I often sleep later than I do at home, sometimes not getting up until nine. That's unusual for me, as in Inverness I'm often up about seven. The quiet is a palpable thing here; sometimes I feel as though I can touch it. In my flat, there is always the noise of traffic or maybe a car alarm going off. Here, there are few sounds. When the weather is dry, the stream that runs a few hundred metres from the door can shrink to a mere trickle, so quiet that I can't hear it when I sit in the bothy watching the passing clouds. At other times, when rain lashes down in the corries high above, the burn can swell with dark water to the point that it threatens to burst from the banks that contain it. Then I can hear it gurgle and babble in the night. Sometimes it sounds like voices calling in the darkness and I have to remind myself that it is only the brook and stop myself from going out with my torch to investigate.

Now that the rut is over and the stags have ceased roaring, the only other sound is the wind rushing headlong down the glen. Sometimes even the wind is quiet and the glen is still. At other times, when gales rush in across the Atlantic, the wind hurls itself in fury against the bothy wall. On those nights, the old building creaks and groans. Its windows rattle and I have to bar the door so the wind can't break in. Sometimes it feels as if I am on a ship being tossed around a vast grassy sea. These, however, are natural sounds and somehow my sleeping brain lets them pass and I lie undisturbed. Perhaps it is simply that I am at peace here.

In my bothy wanderings, I experience a solitude that is rare in the British Isles today. Few folk have the chance to spend a night alone as I have, or even an afternoon. I am sure for many, unused to being away from the hubbub of society, the silence and emptiness of this place would be disconcerting, if not frightening. It is a pity that many of us have lost the ability to be comfortable in a wild place on our own, because this is after all where we came from. How many people live their lives rarely leaving the concrete we have created? Days can pass here when I see no figures on the small track that leads up the glen and around the foot of the ridge to my door. I have to confess, on my long writing days, I often glance up at the track hoping to see a distant figure approaching.

***

One of the great gaps in my experiences of wildlife, and I am beginning to realise that there are a great many, is that I have never seen a whale. Maybe I have fallen into the trap of letting David Attenborough be my eyes and ears. I am a creature of the land, and almost all of my journeys into our hills and foreign places have been landlocked. I have decided that this is something I should rectify and so I am driving over to the Isle of Mull to take part in one of the few remaining wildlife cruises of the year. Soon the seas off Mull will be swept by ocean gales and far too treacherous to be explored by tourist boats. I'm not by nature, as you may have guessed, a gregarious person, and I have an intense dislike of being organised as part of a group. The fact is, however, if I want to have first-hand experience of one of the greatest mammals on the planet, since I don't own a boat and wouldn't know what to do with one if I did, I'm going to have to buy a ticket and do what I'm told for once.

Mull has always been a magical place for me: there is a sense of wildness here. It is green and verdant, its woodlands giving it a very different atmosphere from other west-coast islands where trees are often sparse. Although Mull is a popular island, its sheer size means that it never feels crowded. It's a forty-mile drive from the ferry on the east of the island to its westernmost tip at Fionnphort, the small village that is the departure point for the short ferry crossing to Iona. It's perhaps important to explain what a village is here. The word 'village' in England conjures for me images of *The Archers*' Ambridge: a significant if small hub of population with a church, a shop and several hundred people making regular visits to The Bull and having fetes on the green. On the west coast of Scotland, villages are on nothing like such a grand scale. If there's a shop and three houses, that's a village. If there's a hotel, that's an added bonus. In fact, not too long ago, if there was a phone box, you were pretty close to village status. Villages contrast with everywhere else, where there's basically nothing much. On the west coast, villages are places where you might actually meet someone else.

Mull has become famous for its sea eagles, the superstars of wildlife reintroduction in the UK. One thing that Mull has been able to do is demonstrate how wildlife tourism can have a big impact on rural economies. Sea eagles, also known as white-tailed eagles, were hunted to extinction in the 1800s and reintroduced to the west coast of Scotland in the 1970s. They are the fourth-biggest eagle in the world, with a wingspan of almost two and a half metres. They are formidable hunters, with their main prey being fish, although they will take mountain hares, rabbits and the occasional lamb. I've seen sea eagles several times, mostly high in the sky and at some distance, and one of my dreams is to see one actually hunting its prey.

The eagles have become important to the Isle of Mull economy. Studies suggest that almost a quarter of the island's visitors are drawn to Mull because of the presence of these magnificent creatures. It's estimated that the sea eagles bring around five million pounds into the economy on an annual basis and support over a hundred jobs. That's a very significant figure on an island whose population is a mere 2,800. The significance of this statistic goes beyond the island itself. It demonstrates that wildlife tourism could offer important financial rewards to remote areas of the Highlands and perhaps the opportunity to move away from traditional sources of income based on exploiting the wildlife of the area.

Tobermory is the island capital. Its main street runs along the side of the harbour and is a distinctive row of brightly painted houses in blue, red and yellow. The village has a relatively short history and was established in 1788 by the British Fisheries Society, who wanted a safe harbour on the west coast where fish could be landed and processed. Before the village was constructed, there was only a steep cliff face that led down to a small shingle beach. I've come to join a whale-watching cruise that sails from this sheltered harbour.

Around twenty of us are gathered on the dockside, mostly retired folk like me. We are all here to catch one of the last sealife cruises of the season. Soon the autumn gales will arrive and the seas will be too rough for the handful of small boats that take folk out into the Sound of Mull. I'm here hoping to see a whale. I must have watched hundreds on documentaries but I've never seen one in the flesh. That's something I want to remedy this winter and if I am to do it this season, time is very short. Most of the other folk are well equipped with serious long-lens cameras and professional-style binoculars. I have only the stubby, cheap binoculars I take hillwalking. I look longingly at their expensive equipment. It's a terrible thing, at my age, to experience lens envy. They are all dressed in green; I am wearing my red Páramo. I feel a fraud amongst them.

We climb on board the smart blue and white boat which has a small cabin for shelter and an observation deck above. Inside the cabin, we get a briefing from a suntanned young man: 'Try and keep three points of contact with the boat at all times.'

I can't think why he'd need to say that, as the water in the harbour is silky smooth. He beams enthusiastically at us all, showing his perfect teeth. I suspect that he's never so much as looked at a gooey toffee delight in his entire life. Then he gives a short talk on the identification of seabirds. I nod sagely as if I am an expert and all this is wasted on me, still trying to show the ornithological septuagenarians that I'm their equal.

He grins again and I'm almost blinded by the glare from his molars. 'Let's have a little quiz. I'll hold up a photo of a bird. You write down what it is.'

The green-clad lens-carriers around me greet this suggestion with unbridled enthusiasm. I am less keen to expose the extent of my ignorance. He shows the first photo. At least I recognise that bird. Black guillemot, I scribble down. The next bird he shows is obviously a great skua. I'm doing better than I expected.

Everyone around me is laughing at how easy the questions are. Perhaps my time spent poring over my bird book hasn't been wasted after all. After the final bird has been displayed, we each hand our checklists to the person next to us to mark. I give mine to a tall man from Huddersfield who has not one but two cameras around his neck, and one of those waistcoats with lots of pockets to keep all his bits and pieces in. He marks my sheet and hands it back to me. I get one hundred per cent … one hundred per cent wrong, that is. The birdwatcher from Huddersfield edges away from me, desperately avoiding eye contact. Perhaps I need to look at that book more.

The small vessel takes us out of the harbour to where the ragged landscape rakes the sea with jagged, rocky fingers. We pass the point of Ardnamurchan, a spectacular spike of land reaching out into the sea, topped by the white tower of a lighthouse. In places, the land slopes gently into the water, giving birth to hidden bays and small remote beaches. In other places, sheer cliffs sweep down from the land, speckled white with seabirds. Hidden beneath the waves, the cliffs plunge deep into the sea, creating dark, silent refuges for the sealife. Now the sea is becoming noticeably rougher. I am forced to cling on to rails and ladders to stop myself being hurled across the boat. I'm not used to being tossed about, being of a fairly solid stature; I like to be anchored to the surface of the earth. I find this unstable world unnerving. The quizmaster grins happily and strolls about the deck as if it were flat calm. I lunge from one railing to another, uncertain as to whether the next wave will throw me overboard or send me crashing into one of the many solid bolts and bits of iron that I face at every turn.

The rest of the passengers seem used to this and are happily clicking away with their cannon-sized cameras in every direction.

The crewman who led the quiz sits down opposite me and tells me his name is Andy. 'We're not just sailing about randomly,' he explains. 'We are following underwater troughs and hills. These are the best places to find marine life because of the nutrients among them.'

There is a whole hidden landscape beneath the waves that I have never thought about. 'Do you think we'll see whales?'

He grins and flashes his white teeth at me. 'Well, it's a bit late in the season, but we might get lucky.'

As Andy is speaking, a stocky young man with a shock of curly ginger hair emerges from the small wheelhouse of the boat and looks at him quizzically. 'What do you mean, Andy? We always get lucky.'

Andy nods. 'Well, maybe not always, Steve, but usually something turns up.'

They are both laughing together when Steve catches sight of something and points with excitement. 'Sea eagle!'

The boat erupts into life as camera lenses turn to follow his finger. The eagle is sweeping low across the sea, perhaps sixty metres from the boat. It is hunting, its wings beating slowly and rhythmically as it scours the water for prey. Gulls wheel noisily about it, trying to harass the raptor, but it dwarfs them in size and ignores their attempts at distraction. Most of the sightings I have had of these great birds have been of them soaring thousands of feet above glens and hilltops. Then it is hard to grasp their size as they glide effortlessly on the high currents of air. Here, so close to the boat, the bird's colossal wingspan is awe-inspiring. The eagle turns in a great arc, almost skimming the water with one wing tip while the other cuts through the air. Then it straightens up and begins to descend. It reminds me of a Second World War bomber coming in to land. Then it strikes, taking a fish from the water with one precise, perfectly timed lunge with its talons. I catch a sight of the fish's silver flanks as it struggles to get free. There is no escape from an eagle's talons. The bird heads for the coast, a few miles away, to devour its meal.

Although that's the closest I've been to a sea eagle for many years, I did have a very close encounter with one of these great birds in the early seventies. I was climbing a hill one summer in Kintail and was scaling the last few feet to the sharp-pointed summit when I was suddenly aware of a great shape heading through the air towards me. I ducked down instinctively as the huge eagle banked sharply, missing me by only a few feet. I remember hearing the roar of air through its wing feathers as it braked and turned. I suspect the eagle had been intending to land on the summit when my head, complete with bobble hat, popped up unexpectedly. I don't know if the eagle got a fright, but I certainly did.

As we get clear of the headland of Ardnamurchan, the sea becomes calmer and the islands of Muck and Eigg come into view, with Rum in the distance rising steeply from the sea. These are the summits of underwater mountains rising up from the seabed. Now that the chances of being hurled into the ocean have diminished, I decide to attempt climbing the steep steps to the upper deck that is exposed to the full vigour of the weather. There is probably some technical term for this area such as poop deck, but I've already embarrassed myself in the bird quiz.

I'm surprised to see that there are a few stalwarts up here. They have cameras strapped to them and are wedged into various corners of the railings, and have obviously survived here during our earlier rounding of the horn.

Alex, an older, weather-beaten man muffled in multiple layers of oilskins, sits so firmly crammed into a corner above the small wheelhouse that he could be part of the boat. He peers out to sea, scouring the surface of the water for signs of whales, dolphins or birdlife. He is the spotter whose job it is to shout out directions to the skipper. He sees something and calls down to change course.

'What have you seen?' I ask.

His eyes are glued to his binoculars. 'I can see the birds circling. It might be a bait ball. If we are really lucky, we might get to see a minke whale coming up underneath them, breaching to feed.'

I suppose it's obvious. Where you get fish you get seabirds, so if you can find birds there's a chance of seeing whales. Obvious, but not something I'd thought of before. We get to chatting about how things have changed.

Alex has grave worries about the future of the oceans. 'There's been a forty per cent drop in the number of plankton in the sea,' he tells me. 'You lose the plankton and you lose the fish and the rest of the aquatic life.'

So, just as insect numbers have fallen on the land, plankton – the sea's version of insects – have also declined. Everywhere you go, nature is under siege.

Alex shakes his head sadly. 'You can see it in the basking sharks. You only see two or three of them now.'

'I have seen basking sharks before, off the Isle of Arran, cruising in Lamlash Bay with their huge dorsal fins rising out of the water,' I tell him.

Alex settles himself down out of the spray beside the wheelhouse, never taking his eyes from the sea and the circling birds. 'They say in the old days, you could walk from Ardnamurchan to Mull on the backs of basking sharks.'

I know he doesn't mean that literally, but he's trying to convey to me just how plentiful the seas were. 'That's incredible. Were there really that many?'

Alex nods. 'I used to work surveying the sharks. We'd sit up in the lighthouse in Ardnamurchan and the sea would be covered in them. Not just on one level but layer upon layer. All the way across to Mull.'

'I read somewhere that scallop-dredging is destroying the seabed here.'

Alex nods sadly. 'Aye, it's illegal but it still goes on and hardly anyone gets prosecuted. It's a shame what's happening to the seas. We'll never get it back again.'

Before I can answer, Alex spots something and calls to the skipper to change course again. He points to the horizon and suddenly I can see hundreds of grey bodies leaping out of the sea and plunging back in. In moments, the boat is surrounded by dolphins – sleek, powerful bodies in their hundreds. The dolphins are riding our bow wave, leaping and playing all around us. It's as if we have been surrounded by a marine wolf pack.

Alex must have seen this sight a hundred times, yet still he is straining with his camera, trying to get a good shot of these playful creatures. I have seen dolphins before, but only ever from the shore. Here, as they swim in the bow wave of our boat, I can sense the intelligence of these animals and marvel at a creature that displays such ease and mastery in its environment. They cut through the water like muscular torpedoes and I can hear the clicks they make to communicate. There is no mistaking their sheer joy.

'Whale!' Alex cries, his face alive with excitement.

I turn, but the whale has slipped below the surface before I can glimpse it. This happens half a dozen more times. Each time, the whale vanishes before I can catch sight of it. After an hour or so, we break company with the dolphins and head back to the shelter of the harbour of Tobermory.

I haven't seen a whale this trip. Somehow I am glad, as I know that it is better to achieve an ambition through effort than to have it handed to you. I still have the whale to look forward to. I am happy enough to have seen the dolphins and had the privilege of watching the powerful eagle sweep down and take its prey from the sea. I've seen all this and something else too: I've come to realise that there is so much to explore in my Highland home, so many places to see and people to meet. I know now that there is an unseen world waiting for me; all I have to do is take the time to look.

NOVEMBER

This is the far north. In winter, it is a place of short days and low slanting light, timeless quiet and huge empty skies. Rock-clad mountains erupt from endless windswept moorlands and lonely lochs. Standing among the hills and fjords of Sutherland, I can at last turn my face to the wind and hear the bones of the land calling.

It is November. The month begins in the mellow gold tints of the fading year, where echoes of summer can still be heard. It will end in the dark shadow of winter, as the landscape buries its head to endure the cold months. I have driven north from my home in Inverness and arrived at the Highland village of Lairg, which sits at the head of Loch Shin. When I step out of the car, there is a bite to the wind that stings my cheeks and whispers of winter. Lairg is a frontier town: there is the sense of being at the edge of something, about to cross over the border.

This is the last village I will pass through on my journey north. In the centre of the village, where the waters of the loch lap the roadside, there is a white duck house on a tiny island. It's the only duck house I've ever seen and is known locally as the Broons' house. There's a legend that the house was built by a poacher who was gifted the land by the local laird in return for being taught how to make whisky. It turns out that this is a complete myth and the wee house is actually a leftover from a village fete about twenty years ago. I like the poacher story better.

I call in to the village shop for a few of the things I forgot to pack. The essentials of bothy life: half a dozen candles, a pint of milk and some matches. I already have matches and it's likely that there will be some in the bothy, but I always feel the need to buy just one more box. What if I leave them in the car or a box gets wet? I'm haunted by the fear that I will have no means of lighting the bothy fire or starting up my gas stove. Without matches, I would pass a long, dark, dull night. So I pick up yet another box, just in case the three I already have get damp.

I leave Lairg and drive north; already the day is fading and the shadows are lengthening. I pass the tip of Loch Shin before heading into the empty lands of the north. At seventeen miles in length, Loch Shin may not quite rival Loch Ness in sheer size, but it has a remoteness far exceeding its more famous sister. The shores of Loch Ness are populous, with my home town of Inverness at the eastern end, Fort Augustus to the west and a busy road up its northern shore. Loch Shin is different. In the winter, you can drive alongside its entire length and meet only a handful of cars. On a dark winter's night when the wind carries snow in from the north, it is definitely not the place to break down.

I take the road north and drive past the Crask Inn, which stands alone beside the single-track road and can surely claim the title of the most isolated pub in Britain. I drive on through the empty hills, meeting only two or three other vehicles, and pass through the tiny settlement of Altnaharra. It has a hotel, a handful of houses and a small primary school, and is the only settlement for miles. Its one claim to fame is that it holds the joint record for the lowest temperature ever recorded in Britain, -27.2° Celsius in December 1995. This is surprising given that Altnaharra is only a little over twenty kilometres from the sea.

In summer, Sutherland's roads are clogged with camper vans and convoys of motorcyclists following the North Coast 500, a route invented by marketing men to tempt the followers of signposts north. In late September, someone at head office throws a switch and the motorhomes turn and head back south, and Sutherland falls quiet once more. In November, I drive the remaining twenty minutes to Loch Loyal on empty roads.

It's fully dark by the time I leave the car and set off with my pack to cross the narrow strip of land between Loch Loyal and Loch Craggie. In the blackness, sound becomes the pre-eminent sense. Walking along the shore of the loch, I am acutely aware of my boots crunching on the gravel and the waves breaking on the water's edge. The moon is shrouded in cloud, and only occasionally does it catch the water in its pale light. Walking into the night, it takes me only moments to leave the world of tarmac and men and step into a silent place where the only sounds are my footsteps and the sighing wind. I am the only human being for at least five miles in any direction. There are no lights showing and I can see little beyond the arc of my head torch save the odd glint of the moon reflected on the black surface of the loch. I head away from the water's edge, and can feel the craggy hill looming up to my right more than I can see it. From somewhere in the blackness, an owl hoots. Instinctively, I turn towards the sound and pick out a pair of eyes reflecting bright in my head torch beam.

'Hi, how's it going, pal?' I'm glad not to be alone in the darkness and even this lone night bird feels like company.

The path rises steeply for a few metres and then levels off. Soon I am out of the shelter of the woods. The wind picks up and it begins to spot with rain. The small shelter of Achnanclach bothy is close now. I turn up the power on my head torch and spot the bothy hunkered down in a dip, like a child playing hide and seek.

Pushing open the wooden door, I'm glad of my precious matches. They light the candles and push the darkness into the corners of the room. They light the kindling too, and soon yellow fingers of warmth begin to dance amongst the coal. They light my gas stove, and soon I have water boiling in the pan. With the water, I make instant mashed potato, which I enjoy along with a tin of stewing steak. It's typical of the simple meals I take to bothies. It's quick to make and sustaining, perfect for a cold bothy night. All this from a few wee sticks with tiny red heads. Someone should tell Ray Mears about matches – they'd save him hours of that stick-rubbing business.

After my meal, I settle down in front of the fire, enjoying the warmth and looking forward to a whisky or two. Outside, the rain is beating against the windows, and a traveller on the road a few miles distant across the loch might spot a lonely light on the black hillside and wonder whose house it is and what makes them live in such a place. Perhaps they would pity me, perhaps envy – I cannot say, and care even less. As the firelight dances, my contentment grows, and the rising wind beyond the stone walls of the bothy serves only to fuel the satisfaction I feel. On nights like these, I try not to think too much. We spend our lives worrying about the future or regretting the past. We forget that all it takes to be happy is a full stomach, a warm hearth and the sense to let the world be, just for a little while.

The morning is still when I take the short path to return to my car. Last night's rain has cleared, leaving only pools to mark its passage. The loch stretches endlessly away, its dark waters leading my eyes towards the distant snow-dusted hills. The loch is mirror-still, as if its water has turned to glycerine. Beyond the shore, my boots thud across the wooden bridge over the tiny river that cuts through the isthmus and joins two lochs together. The river is only a hundred metres long here, but I can never resist pausing in such a place and looking deep into the water; my teenage years spent angling have left me with an unending curiosity about what lives beneath the surface.

A narrow strip of sand, less than a metre across, borders the river water and separates it from the shingle. I scan the sand for signs of passing wildlife. A flat beach is like a book recording the signatures of any man or creature until the next wave erases them or the rain and weather wipe the sand clean. When I was small, my father pointed out to me the signs of paths made by mice and other animals, tiny roads through grass or leaf mould created by the passage of a hundred micro feet. Once seen, I could not forget such tracks, and sought them out whenever I could. Even now, I always take a moment to see what tales the marks in the ground can tell. Sometimes I see a badger's paw print, often deer, and now and then the rounded imprint of the toes of a pine marten show in the mud. Any hillwalker will soon learn to distinguish between a path made by the feet of people that is worth following to a summit or perhaps back to the security of a road, and the path of a deer that will take the walker off into the unknown. Learning the signs of animals is little different.

I check one side of the bridge, but the sand is blank. I check the other side, where the sand narrows to meet the shingle less than half a metre away. At first I see nothing written on the fine, granular surface of the beach. Then I see what I am looking for. Just at the water's edge, there is a sign the wind has missed. It is the paw mark of an otter, made as he stepped from the water. As he rose from the swirling river, his left foot made an impression, pushing a pebble into the sand. There is a slight mark where the claws of his right paw grazed the surface and raised a small ridge.

A faint trace of a moment in time. The prints remain sharp in the sand. If the rain returns, they will vanish by the afternoon; if not, this light wind will take them away in a day or so. He was here in the grey light of dawn this morning. He touched the sand for only a fleeting moment, but the mark he left is all I need to see him. He was a dog otter, a big male – I know that from the size of his paw print. He stepped tentatively from the water, his body wet and sleekit, whiskers twitching as he sensed the air. Thinking himself safe, he moved nimbly on to the gravel and searched for a moment for some small prey; finding nothing, he moved on. I follow him. I cannot see where he walked on the gravel, as he was too light to leave a mark here, but I don't need to find a track; I know where he was going. He won't have strayed far from the water. I follow him for ten or twenty metres until the gravel gives way to sand at the edge of the loch. Here he moved quickly, perhaps feeling vulnerable on the open stretch of sand. He left his paw marks, even though they are fading fast where the beach is exposed to the wind. I can make out the sweep of his tail through the sand as he entered the water once more.

My boot prints mark my passage beside the otter's. He and I are brothers in this wild place. Perhaps, somewhere out in the loch, a small brown head bobs on the surface of the water and watches me as I follow the trace of his footsteps. We travel together, our worlds separated by a thin meniscus.

An hour and a half later, I am walking with my coal-laden rucksack towards another refuge, my mind submerged in this remote landscape. The moorland stretches away ahead of me, and a range of low hills stands between me and the coast I am heading for. This is an unusual walk for me. I rarely walk downhill to a bothy. Winter has arrived early here. The wind is keen and numbs my cheeks. In the distance, high hills rise jagged-toothed from the peatlands. Already they are swathed in snow and the sky is filled with edifices of dark clouds jostling for space in the emptiness. I am returning to a secret place. The journey here is never easy; the terrain is rough and the weather rarely kind.

The road feels a long way off now and the security of my little Skoda is a memory. I am alone here, an hour's walk from the tarmac. I have only what my rucksack can hold, can only travel where my boots can take me, but that's okay. I'm going nowhere, and if nowhere is a place, then I am already there. There are no paths here and few signs of the passage of other travellers. No boot prints or any evidence of human footfall. Even the deer tracks are scant and fade to nothing after a few metres. This is an untrodden land whose very remoteness has drawn me here. I weave between tussocks of grass, search the endless bog for dry footfall and, finding none, move on with my feet squelching with every step.

Every few hundred metres, I come to a stream. Some I can ford by stepping on convenient stones; others are not so easy. I am teetering on the brink of a fast-running mini-river. I hop on to a few small boulders and then reach the middle, where I must leap for the bank. My mind says leap, but my body does a semi-controlled stumble into the water and on to the steep bank on the far side. I land face down and the peat bank I fall on to decides, after years of waiting, that this is the moment to collapse. I am thrown back into the water. I scramble and kick, arms flailing, boots splashing. To the casual observer, I look like a tortoise having a fit. Somehow, I end up vertical once again, grass in my mouth and my rucksack soaked and hanging off me at an odd angle. I'm gasping from the exertion and swearing at the same time. You would think that in all my years of hillwalking I'd have mastered stream crossings. Sadly, that does not seem to be the case, but I have learnt to accept that once in a while I will end up flat on my face. Such zen-like acceptance takes years of practice.

The bothy sits on the coast a few hundred metres from the sea. It is stone-built, windproof and watertight, and has changed little since I was last here the previous summer. Because of its environmental sensitivity, I am not going to name it. With a little detective work and an OS map, it is probably fairly easy to deduce where I am, but it is not maintained by the Mountain Bothy Association and therefore not public, so I am not going to reveal its exact whereabouts. I hope you will forgive my reluctance to share such places, but they are rare and precious.

The casual wanderer is unlikely to stumble across it, but if you are determined, you will get there.

The bothy is a simple stone house sitting among the heather with its face turned towards the sea. Inside, it is crammed with artefacts gathered from the shoreline by previous occupants. There are pieces of weirdly shaped driftwood, creel floats and a multitude of shells. To my amazement, I find it supplied with tins of food. A note in the bothy book tells me that a party of men picking winkles on the shore had used the bothy for several days prior to my arrival. They had come in by boat and took pride in writing in the book how they had tidied the bothy before leaving. The place was very tidy, but I was further amazed to find a large bin bag full of bottles and tins that they had left behind. Since they had a boat, it would have been so easy to take it out – unless, of course, their rubbish bag was capable of sprouting legs and walking out of its own accord. Perhaps they thought it would be taken away at the next bin collection day, which is, of course, never. I set up my stove, light the coal I carried across the miles of moor and settle in to my bothy night.

The following morning, a cold wind sweeps past the door, carrying with it a suggestion of rain. Below the bothy, the land stretches for a quarter of a mile or so before it tumbles into the sea. Some two miles to the north, there are great cliffs that hold within them sea caves the size of cathedrals, the mark of thousands of years of rain trickling down through the limestone and the passage of a million tides. Here and there are small, boulder-strewn beaches and it is these that I have come to this northern shore to visit. Here I might find the prize I seek.

Between the bothy and the sea, for a distance of 200 metres, the earth shows a series of man-made ridges. These are about 2 metres across, 150 metres long and 30 centimetres high. These ridges, known as runrigs, are a clear sign that the bothy was once a home to a family who farmed the land, for at least a few generations. Runrigs came about because the depth of Highland soil is frequently so shallow that it would struggle to support crops. By piling up the earth in rows, enough soil depth could be created to grow potatoes and other vegetables. Higher on the hill, I had passed several distinct lines in the peat bog where the people who lived here had cut rectangles of peat a little larger than a man's hand and stacked them to dry so they could be used as fuel for the fire.

It must have been a harsh life. Lashed by endless wind and rain, almost literally clinging to life on the edge of the Atlantic. This place feels so empty now, yet children were born here, folk grew old and passed away their years. Perhaps if I had stood here 150 years ago, a red-haired girl would have come running barefoot down the hill laughing, with her collie dog at her heels. Surely these hills remember a time when the chimney of the house always smoked, when hands dug into the dirt and brought forth life. There is a memory in this place that speaks to me through the sky and the stones.

I reach the first shingle beach where the waves sweep in. There are small pools here, protected from the full force of the winter storms by a wall of rock. Ideal shelter, one would think, but the beach is deserted, with only two dark cormorants drying their wings in the sun high on the rocky cliff. I search two more beaches and they are deserted as well. There is only one beach left to try. Perhaps I have come all this way for nothing. Round the next headland, I follow a small stream into a narrow ravine where it meets the sea. I turn a corner and a rock moves. I back away quickly, not wanting to disturb anything. I crawl to the cliff edge that overlooks the beach and find myself face down in a pile of sheep poo. I wonder if David Attenborough ever lay face down in sheep shit.

Peering through my binoculars, I see other rocks moving on the beach, more and more coming into focus. These are seals nurturing their young. They are so well camouflaged that they appear to be boulders. I can see around twenty of them lying on the beach with their young pups beside them. This is what I have walked all this way to see. For just a few weeks in November, seals have their pups on these rugged shores. There is a small sandy beach here, no bigger than a football pitch. Protected from the sea by bands of rock are small pools where the young seals can gain their first experience of the water in safety. The grey dappled skin of the adult seals perfectly matches the boulders they are sitting beside. The pups are lighter, almost white in their juvenile fur.

Before this, my only sighting of seals had been heads bobbing in the water, watching me with curiosity as I walked along remote beaches, or sometimes basking in the sun on distant sandbanks surrounded by the sea. This close and out of the water they are much bigger than I had imagined. Their bulbous bodies are surrounded by great rolls of fat to protect them from the cold seawater. On land, they struggle to move and can only make progress in ungainly short waddles as they try to move their bulk across the sand. Not much happens in the seal colony. The pups lie beside their mothers, contented and safe. Occasionally an adult waves a flipper and the young come close to their mothers to suckle. I am watching a scene that cannot have changed since the ice retreated 10,000 years ago. It is one of the few truly wild sights that still exist on these crowded islands.

Even in these remote waters, the seals are not safe from the hand of man. Every year, hundreds are shot by fish farmers protecting their stock. For this reason, American companies are banning the importation of Scottish farmed salmon. The salmon-farming industry that has sprung up around the shores of Scotland provides valuable jobs for small isolated communities, but is far from the natural and sustainable industry one might imagine. Salmon-farming has grown dramatically over the last few decades and there are now over 200 fish farms producing more than 150,000 tonnes of salmon a year in a billion-pound industry.

Salmon do not naturally live in netted pens, and fish farmers are constantly in a battle trying to keep at bay epidemics of parasitic lice which literally eat the salmon alive. The amount of farmed salmon is now dropping due to infestations and disease. Pesticides, waste food and faeces pollute the sea lochs that support the salmon farms, causing huge environmental damage. I never eat farmed salmon and I am sure many people would avoid it if they knew the damage the farming process causes and the undoubted suffering of the fish themselves, let alone the casualties in the seal population.

I slip silently away from the cliff edge, rendered pungent by the sheep poo. I'm glad to have seen these seal mothers with their pups. I have achieved one of the goals I set myself this winter and have at least found one place where wildlife has managed to cling on to survival. I feel a deep sense of satisfaction on the long walk up from the coast and back to my car.

***

After my journey to Sutherland, the weather turns angry. For a week, high winds sweep in from the east, bringing snow to the high tops.

I'm sitting in my flat one Thursday evening when the phone rings. 'What are you doing this weekend?' It's my friend Martin, who I walked the Pennine Way with forty years ago.

His call takes me by surprise. 'Nothing planned yet. Why?'

'I thought I might come up. We could do a hill, a bothy maybe?' Martin sounds enthusiastic in his Blackpool home.

'I've not seen the weather forecast.' I hadn't been expecting to hear from Martin until the new year and our traditional outing in February.

'I have,' he explains. 'It's getting calmer and colder. Might be okay.'

I think for a moment. The days are shorter now and it's hard to carry in enough fuel for a bothy, but there is a mountaineering hut I know. 'How about Inver, at Achnasheen? Plenty of options for the hills there.'

'Achnasheen!' Martin is excited. 'I could get the train there.'

He doesn't drive and has been a devotee of the rail network all his life. He'll be coming to Inverness on the overnight train and is delighted by the thought that after twelve hours on a train he'll be able to add another hour or so to the remote Highland station. I'll be driving there and he could easily come with me, but that would waste the opportunity to enjoy rattling along the Kyle line. The route is acknowledged as one of Britain's most scenic.

Fifty minutes after driving out of Inverness, I am pulling into the station car park at Achnasheen. It is about as small a place as it is possible to imagine. There is obviously a station there, but before you begin to conjure images of ticket offices, bookshops and cafes, I'd better explain that it is only an unmanned platform. There are about half a dozen houses. There used to be a shop, but that closed down; there was a petrol station once, but it closed too. There was even a hotel, where Martin and our friend Joe and I took refuge from the winter weather one Hogmanay, but that burnt down a long time ago.

Achnasheen is Gaelic for 'field of rain', and it more than lives up to its name for large parts of the year. Gaelic names are frequently little more than literal descriptions, no more than 'big pile of grey stones' or 'odd-shaped hill.' The people who populated these hills saw them more as obstacles than romantic places of escape.

Despite its small size in terms of population, Achnasheen remains a significant junction. Not only does the rail line to the west coast pass through it, but it is also where the road from Inverness splits. The right branch heads on towards Torridon and Gairloch, and the left branch heads to the villages of Lochcarron and Kyle of Lochalsh, which is the gateway to Skye. So grand is this junction that it possesses a very rare feature in the Highlands: it has a roundabout. Sat incongruously amongst these sparsely populated hills is an urban-style roundabout. Although the level of traffic here does rise in the summer, at no point can it ever be described as busy. The roundabout is a planner's work of fantasy, and almost certainly the result of an EU grant being available for supporting rural infrastructure, so if the money is there for a roundabout, you might as well have one, even if it's obviously not needed. Who knows, it might come in handy one day.

A chill wind greets me as I step on to the platform and I retreat into the folds of my duvet jacket. The platform is empty. Above me, the hills are glazed white, so we might yet get a chance to climb a winter hill, even if it is early in the season. Two minutes later, I see the twin carriages of the diesel coming into view. A minute after that, the tall figure of Martin steps beaming off the carriage, dressed in his customary woollen trousers, tartan shirt and 1970s balaclava.

'Did you have a good journey?' I ask by way of greeting.

'They tried to take my ice axe off me,' Martin announces in astonishment as we walk.

'Really?'

'The jumped-up little guard on the train at Preston said it was an offensive weapon and I couldn't have it with me.'

The ice axe in question has a steel spike on one end and the other end is garnished with a cutting blade and an inclined dagger. A determined assailant could turn carriage B of the 8.40 from Preston to Carlisle into a bloodbath before the refreshments trolley had even made its first run. Martin is obviously unaware of this.

'I told him this axe is irreplaceable. You're not taking it anywhere.' Martin is correct: they stopped making ice axes like his thirty years ago.

'What happened?' The outcome of Martin versus authority is always interesting.

Martin shrugged his shoulders in frustration. 'He said it was his train. I told him it was my ice axe and he better take my woolly balaclava in case I strangled someone with it.'

'Stalemate! What happened then?'

'The police turned up.'

'My God! What did they do?' Martin's confrontations with authority normally stop short of the constabulary. Perhaps threatening to murder someone with a woollen hat raised the alert level.

'They didn't seem too bothered, just let me keep it.' He shrugs and shoulders his colossal rucksack as we both head for my car.

It's gratifying to learn that the rail police are able to distinguish between an eccentric middle-aged mountaineer and a member of a terrorist organisation. There ought to be more people like Martin about, able to demonstrate that a determined individual can still overcome the gargantuan edifices of modern bureaucracy.

Martin 1, Rail authorities 0.

It is a short, two-mile drive from Achnasheen station to the path leading to Inver hut. It's almost dark now and the wind has dropped. Inver is separated from the road by a small river and a formidable bog that runs between two lochs. It is prone to flooding and there is a danger that the folk in the hut could be marooned if the water rises. To solve this problem, a long wooden walkway has been constructed a metre and a half from the surface of the bog.

Our boots crunch on the sparkling hoar frost coating the boardwalk as we lug our rucksacks towards the hut. Today the air is still, but when the wind is high this short walk to the hut can be very different. Some freakish feature in the hills channels a ferocious wind up the glen. The boardwalk raises visitors out of the bog but, in doing so, it lifts them into the full force of the wind. This odd venturi effect creates surprisingly powerful winds to such an extent that on a couple of occasions I've been forced off the walkway and had to wade the bog for fear of being blown off.

'What's this hut like?' Martin asks as he follows me across the boards.

'Oh it's got a gas supply, running water, electric lights. Luxury!'

'A sort of super bothy, then?' Martin laughs.

'I suppose it is.' I laugh a little too loudly and Martin senses something.

He sighs. 'Is there a catch?'

'Yes. It's the coldest place on earth,' I admit. 'Colder than any bothy I know.'

Martin is always positive. 'Ah well, we've plenty of coal.'

Inver mountaineering hut is a low stone cottage less than 300 metres from the road. It once lay roofless, derelict and forgotten, but it was slowly brought back from the grave by the Jacobites Mountaineering Club. The work was carried out by club members working at weekends and took twenty years to complete.

Half an hour later, Martin and I are hunched over a roaring fire. I'm swathed in my down duvet and Martin is wearing two of his 1974-style woollen pullovers. The flames of the fire are leaping up into the chimney, but the temperature of the room has not changed from near zero.

Martin speaks, his breath misting the air. 'Is it always this cold?'

I blow on my hands, trying to get some life into them. 'Well, nobody's been here for three weeks. It'll warm up eventually.' I try to sound encouraging.

Martin groans and heads off to use the bathroom. 'Oh God! Come and look at this.'

Fearing some terrible plumbing disaster, I leap up and find him staring down in to the toilet bowl. There, floating in the water, is the corpse of a drowned mouse.

Martin sighs. 'Poor thing. Must have slipped on the rim and fallen in. Couldn't get out.'

'Nah. I don't think that's what happened.' I turn to Martin and fix him with my most doleful of looks. 'I think it couldn't take the cold and starvation of this place. Suicide.'

Martin nods and depresses the toilet handle and we commit the body to the drains.

An hour later, the fire has begun to make an impact on the room. I risk unzipping my duvet and Martin takes off one of his pullovers. 'You don't normally come up in November. What inspired this?'

Martin is quiet for a moment. 'I don't know how long I'll be able to do this: get up the hills. I decided I'm going to make the most of it while I can.'

Martin has reminded me that neither of us are getting younger. I poke the fire and a shower of sparks heads up the chimney while I think how to answer him. 'Well, let's make the most of tomorrow, then.'

We awake to a frozen stillness. With the fire long since dead, Inver has become frigid once again. After a hasty breakfast made with our breath smoking the air, we pack our rucksacks and head out just as the short winter day is beginning. The walkway to the road sparkles and crackles as we head away from the hut. The road down to the village of Lochcarron runs between a dozen or so hills. We choose one that looks close to the road and head on up. I'm not selective about which hill I climb; I just relish the chance to leave the world behind me for a few hours and spend a little time with winter.

The long glen is lined with conifers. The yellow and gold flanks of the November hills rise steeply out of the trees, and the last thousand feet of the hills are christened white with early snow. We cross the railway line and begin our way up through the bog. We are lucky: last night's hard frost has turned it to iron. Normally we would be picking our way between deep pools of peat, but today we can step without caution. The grass is jewelled with hoar frost and the hills are transfixed with cold. The burns have been silenced by the ice, and as we climb higher, the hills around me offer a view into a frigid world. As I make my way slowly up the ridge, I can feel my knees creaking. Martin follows a few metres behind until our boots break through the snowline. Now that we are high on the ridge, the glen reveals itself below us, a green ribbon snaking through the frosted hills.

We pause for a minute or two, both blowing hard. It has taken us longer to get here than we expected. These days, everywhere seems further away than it was and the hills have grown steeper.

Martin surveys the last few hundred feet of the hill between us and the summit. 'Looks pretty icy up there.'

I realise he's right – there is an unmistakeable sheen on the surface of the steep snow. 'Better put our crampons on, I suppose.'

I rummage in my rucksack but can't see them. 'Oh bugger, I've forgotten them.'

'You sure? I thought I saw you put them in.'

I look more determinedly and find the bits of ironmongery hidden beneath my spare gloves and sandwich box. 'Ah, here they are.'

Martin laughs. 'Imagine the headlines: famous mountain author falls to his death after novice mistake of leaving his crampons behind.'

The newspapers always like to blame someone. 'Yes, I can see that headline now. I'd never live it down. Even if I was dead.'

I pull my old crampons out of my rucksack but they struggle against me, a jumble of metal spikes and straps, fighting like a frantic spider trying to escape. It takes a few minutes of fumbling in the cold wind to subdue them and get them fixed to my feet. Once attached to my boots, the crampons remember their job and bite through the icy crust with a reassuring crunch. For the first few steps, the metal teeth feel unfamiliar on my feet, but soon my body remembers how to work with these tools and we head slowly upwards into the white.

As we reach the summit, a stiff breeze greets us. The wind is so cold it brings tears to my eyes, even though I am buried beneath my balaclava. When the spindrift eases, I glimpse the white-capped hills stretching far away. We stand gazing at the view.

'Here we are again,' Martin says, between gasps for breath.

I look out across the vista of snow-glazed peaks rolling across to the horizon. 'We've come a long way from Edale.'

Martin laughs, his mind travelling back forty years to our teenage trek up the spine of Britain. 'God, remember those bloody peat bogs?'

My turn to laugh. I am here once more, in winter's embrace. We have stepped into the world of white imagining. Part of me welcomes the cheek-numbing cold, enjoys the spindrift in my eyes, as I remember days on these hills over so many winters. Not that far from here, I shivered in an icy Bearnais bothy, bivvied out in a magical frosty night and wandered further still to the remote Maol-Bhuidhe bothy. We both stand for a minute, recalling our journeys into these fabled places.

Martin turns to me. 'Pork pie?' He thrusts a miniature Melton Mowbray towards me in his Dachsteined hand.

'Yes please.' Reminiscence is a fine thing, but on a cold summit you can't beat a pork pie.

We spend a quiet evening in Inver trying to take the ache out of our legs with the heat of the fire. The following morning, the train pulls out of Achnasheen with a potential axe murderer sitting in carriage B.

***

A week later, I drive south down the A9. After twenty minutes, I turn off the dual carriageway, pass through the distillery village of Tomatin and enter the rolling hills beyond. These are the Monadhliath mountains, an area of gentle rounded hills, bordered by Loch Ness to the north and the villages of Kingussie and Newtonmore to the south. To me, they are the foothills of the Cairngorms, similar to their grander neighbours in their lack of sharp ridges, both topped by a high plateau swept by icy winds. The range slopes gently northwards, towards the great glen where the dark plunging depths of Loch Ness divide the Highlands.

A naturalist friend first told me about this place. When he spoke, his tone was reverential and there was wonder in his eyes. He told me of a wild land rarely roamed by men, a place populated by eagles, otters and wildcats, where the last wolf in Scotland took refuge. That was forty years ago and, although neither of us knew it, the Monadhliath he spoke of then was already a memory. The Monadhliath I found when I began to explore them were broken hills, damaged and desolate. They are subjugated by sporting estates for driven grouse shooting. Their summits are burnt by controlled fires to prevent regrowth and they are criss-crossed by hundreds of miles of Land Rover tracks. Wind farms dominate the landscape. Gates on to the land carry notices warning of shooting parties and make it abundantly clear that hillwalkers are not welcome. Golden eagles mysteriously vanish here, and hidden in the heather are traps set for weasels, stoats and any other predator that might threaten the red grouse population.

I rarely visit these hills, even though they are close to my home. The place stinks of death and I find no beauty in its landscape. When I walk these hills, I am constantly reminded of the damage that has been done to them. I see it in the patchwork of burnt heather, in the scars of bulldozed tracks. It's in the lines of shooting butts strung out across the hills. The peat-blackened streams that run across the landscape would not be so lifeless were it not for the water running off the scoured earth. This is a landscape in pain and I take no comfort in journeying here.

Today is different. Today I am visiting these hills with James Roddie, a wildlife and photography guide who lives on the Black Isle just north of Inverness. He is a tall, lean, dark-haired young man who spends most of his life outdoors. He regularly rises from his bed in the middle of winter nights and climbs hills so he can capture stunning images of the mountains at dawn. James has told me about one estate in the area that is making an attempt to redress the balance of nature and has become a haven for the mountain hare. Our quest is to get up close and personal with these hares, which is not an easy thing.

We pull up in a small car park beside the Findhorn, one of Scotland's great rivers. It is born in the heart of these hills and flows through these high moorlands over fifteen miles until it courses beneath the A9 trunk road. From there, it travels on, finding its way through the rich Moray farmland where barley is grown to keep the whisky stills supplied. At last, it empties into the sea at Findhorn Bay, some sixty-two miles from its source in these hills.

When I step out of the car, ice is building in the eddies of the river, unusual this early in the winter. The low hills around us are part of the Coignafearn estate. Here, deer numbers have been reduced, the practice of burning the heather has been abandoned and no hares have been shot for four years.

The landscape is dusted with snow, the ground is frozen and our footsteps reverberate through the wooden bridge. As we reach the far side, a barn owl glides out from beneath it, silent and ghost-like, its body the model of perfection. It is as though each feather was placed with absolute precision by an engineer of genius. It has a supernatural presence in the air, moving through it without causing the slightest ripple. The owl sweeps away from us in a soft, effortless arc and vanishes behind a small stand of trees.

I turn to James. 'I wonder if he's landed there.'

James is as excited by the chance sighting of this beautiful creature as I am. 'Let's take a look.'

We walk quietly round the trees, hoping to catch a second sighting of the bird, but we are disappointed. He must have woven noiselessly between the trees and out on to the open hillside beyond. Barn owls are relatively common throughout Britain but they are rarely seen. Any creature is a miracle of evolution, even our two-legged kind, but the owl is a creature so perfectly adapted to its airborne existence it is impossible to imagine how it could be improved.

We walk for half an hour, slowly climbing the steep track until we are on the flank of the hill.

'I've seen quite a few hares here before.' James scans the hillside with his camera, searching for our quarry.

I can see nothing on the hillside but heather and patches of snow, but James seems confident. Mountain hares were once common across the whole of the Highlands, but they are now only to be found in any profusion here and in the Cairngorms. I have seen them many times, but never up close. I've only ever spotted them darting from their holts on my approach. At this time of year, the hares are in the process of changing from the dull brown of their summer colouring to the near white of their winter coat. This transition evolved to give the hare greater protection from its principal natural predator, the golden eagle. The rise in global temperatures and consequent lack of snow in the Highland winter means this change can have the opposite effect. Against a dark background, the poor white hare might as well be waving a flag with the words 'Here I am'. Good news for golden eagles; bad news for hares.

Today the hillside is covered with patches of snow interspersed with moss, grass and heather, giving a mottled effect of white and brown. This combination of colours is identical to the hare's; as a result they are virtually invisible. Good news for the hares; bad news for us.

James has a trick up his sleeve. He produces a camera with a lens like a bucket. I exaggerate – a small bucket. He photographs the hillside and then enlarges the image in his viewfinder.

'I'm hoping we can find a sitter,' James explains as he examines the camera's screen. 'Mountain hares have different personalities. Most run off when you approach, but about one in four sits still and lets you get close. Sometimes you can almost touch them.'

I've never seen a hare sit still for more than a moment and I'm dubious we could get that close, even with a ketamine-laced carrot.

'There's a hare,' James says, passing me the camera.

'Oh I see.' I don't, but decide my failing eyesight must be preventing me from seeing the animal and don't want to admit it. 'Actually, I can't.'

James points to an area on the screen of his camera. I can see nothing there. 'Oh yes, now I see it.'

'The best way to get close is to approach them from below, make sure the hare can see us. That way they feel safe. They know they can outrun us and pretty much anything else, so we're no threat.'

I know that a hare being chased by a dog will always run uphill. Because of the hare's long back legs, no dog has a chance of catching it in that direction. That means that a hare looking down on us will feel secure and be less likely to bolt.

James points at the hillside. 'There, beside that rock.'

I follow his finger and in the orbit of my binoculars one of the patches of snow magically turns into a hare. He is pressed down low, conserving heat on this cold day. He watches us with his burnished eyes, ears twitching and nose sensing the air. He is aware of us, but doesn't appear concerned.

James tells me to walk behind him so we can make as small an outline as possible. We move towards the hare slowly, stopping every few feet to let him get used to our presence before we move closer.

James turns and whispers. 'When they get agitated, their eyes pop open wide and they sit up on stiff legs.'

We inch closer. The hare watches us, unconcerned. We are perhaps twenty feet away, closer than I've ever been to a hare. I can see his twitching whiskers and the texture of his fur. He appears to have one layer and then other hairs protruding out, longer than his base layer. There is an unmistakable intelligence in his eyes. He sits and watches, we move, and he watches us again. We are far closer than I thought possible, fifteen feet now. Then, as if the hare sees some hidden signal, his eyes open wide, his ears go up and he sits bolt upright. We both freeze, but he's made his decision. He rises, turns and runs away up the hill. His movement seems fast to us, but he's not doing much more than a stroll for a hare. He's not our sitter.

We climb higher up the track. Below us, the green ribbon of the valley of the Findhorn snakes its way into the heart of the Monadhliath. The ruins of old farmhouses are scattered along its length, showing that this glen was once more populous than it is today. As we move higher on to the shoulder of the hill, we leave the shelter of the glen and the cold wind finds us. There is an icy feel to this place now. We are only a few hundred feet below the white frosted summit of these hills. It is an empty, treeless place. I realise that there could be hundreds of hares sitting watching us and we would never see them, so well adapted is their camouflage. I am beginning to love this kind of walking. We are moving slowly through the landscape, taking time to explore it, looking for the prints of passing creatures in the snow. Here and there are the paw marks of a mountain hare, a fox who passed by last night, a stag who strolled through the landscape. This is much more than simply heading for some appointed high place. Now the land is speaking to me, telling me its secrets, and I am learning to listen.

James spots another hare sitting on the hillside above us. I can see it now – I'm beginning to get my hare-spotting eye in. We follow the same process, moving closer to him and pausing to let him get used to our presence. It takes twenty minutes, but now we are as close to this hare as we were to the first one. James stops, leaving me to crawl forward. I ease my body slowly through the snow, trying to ignore the cold. I am so close to the hare now I can see his whiskers twitching in the wind.

He is dozing. Now and again, he opens his eyes, takes a look at me and then, like a bored TV-watcher on a Sunday afternoon, his eyelids droop and he slumbers again. Less than ten feet from him, I can see what a beautiful creature he is. His fur is luxuriant, his eyes brown and shining. I have to keep reminding myself that this is a wild creature and he is allowing me into his world.

James joins me. I'm delighted to be this close to an animal that I have only seen fleetingly in the past as it fled from me across the moorland. Now he has accepted that I am no risk, the hare sleeps fitfully, sheltered by a low bank of peat. An aircraft cuts noisily through the air above us and the hare starts in alarm.

'He thinks it might be an eagle,' James whispers.

The plane passes and the hare settles and dozes once more. Lying full length in the snow is something the hare is comfortable with, but not something I can do for long in the cold. This is a harsh environment, and survival here rests on a knife-edge. The hare will not want to waste energy in a full-speed sprint, and I know that if I move much nearer, his tolerance of me will break and he will turn on all his power to escape. I force myself to overcome the temptation to get closer, and instead start to retreat. By the time I have slithered away a few feet, the hare is comatose, perhaps dreaming of carrots. I'm glad I didn't disturb him – we shared a few minutes together and he wasn't stressed at all.

James and I sit on the track overlooking a tributary to the Findhorn running in from the east. The far side of that burn is a different estate, the patchwork of muirburn scars evident on the hillside.

As we share a flask of coffee, welcome on this cold day, a memory returns to me. 'When I was in the rescue team, I did my first search in these hills. We were looking for a mail plane that had crashed. We formed a long line of men, maybe twenty of us, and swept the summits.'

James looks up from his coffee. 'When was that?'

I search my memory and realise a long time has passed since that misty September day. 'Must be twenty years ago. I remember the mountain hares that day. They ran in front of us, put up by the line. There were hundreds then, dotted all over the heather.'

There are not so many here now. The population is only one per cent of what it was in the 1950s. These creatures are persecuted by the men in tweed who seek to protect red grouse shoots for their wealthy clients. The keepers here celebrate the slaughter of hares. They kill hundreds in a single shoot. Sometimes they lay out all the hares they have killed in rows and stand grinning beside them while photos are taken. I cannot understand such men. I don't agree with the culling of mountain hares, although I accept that some animals do need to be controlled. Killing as a matter of necessity I can accept, but to revel in death as they do is an alien thing. It is estimated that 26,000 mountain hares are shot every year in the Scottish hills.

There's a theory, and it's only a theory, that mountain hares can spread a disease transmitted by ticks which move between the hare population and the grouse. The louping-ill virus is a disease of the brain and central nervous system. It is present in red grouse and some cattle. Despite the lack of any clear evidence that a high hare population adversely affects grouse numbers, the blood sports community not only kills thousands of mountain hares but seems to revel in the practice. The Scottish government has now recognised that there is no justification in such large-scale culls, but has done nothing to ban the practice. Scottish Natural Heritage have asked estate owners to adhere to a voluntary ban. That's a bit like asking a grizzly bear with a sore head to try to be nicer to people. It is disgraceful that Scotland's wildlife is treated in this way. The problem with such culling is that it happens in out of the way places where few members of the public stray, so it's hidden from view. There is a quiet holocaust taking place in our hills. We all have seen films and read books about post-apocalyptic worlds where the fabric of society has been destroyed by aliens and a few survivors cling to existence in secret corners. For the wildlife of our hills, we are the aliens and the apocalypse has already taken place.

DECEMBER

December in the Highlands is rarely the coldest month of the winter, but it is always the darkest. In the bothy where I am writing these words, it is three thirty in the afternoon and the feeble light of this winter's day is bleeding away. It is already dark beneath the steep crags, and the streams run black-veined down the hills. I set two hurricane lamps down on the table and begin my nightly ritual of filling their tanks with paraffin. The scent takes me back sixty years to boyhood days spent with my father in a fishing hut on the Wirral. In that rickety wooden shelter, the air was filled with the muddy smell of tench, dank-smelling nets drying on low hooks in the wooden roof and steaming mugs of tea. We cooked on an ancient Primus stove that leaked paraffin. The odours of my childhood have etched themselves deep in my memory and instantly take me back through the decades. The images of my early days have faded from my eyes, but the scent of a wet duffel coat or a mist-filled morning propels me back to something deeper than memory.

The cold wind that has been pushing against the rough bothy walls all day subsides with the dimming light, and the twisted glen falls silent. I check the fuel store. There are four bags of coal remaining from the ten that the ghillie brought up the glen on his Argocat five months ago. He won't be making another trip until spring, so I have to save the precious black stuff. At last, the time comes when I can light the fire. This is the highlight of my day. I swing open the heavy iron door and put a match to the kindling. As soon as the flames tickle the slivers of wood, the stove begins to emit a gentle roar. The heart of the bothy is alive now, beating in the chest of black iron. It's an hour or so before the heat of the stove begins to win against the air. Now the cold is in retreat I can at last take off my duvet jacket and woollen hat.

I sit down at the old rough table to write. Only armoured from the cold in layers of warm clothes can I write at all, but the advantages of bothy writing far outweigh its discomfort. The bothy is quiet. Here there is no internet, TV, radio or people to chat to, no distractions, so things get done. There is more to spending time in a bothy than the silence it offers. The longer I spend here, the more I become absorbed by the landscape, become a part of it. My life is tuned to the rhythm of these short December days; my existence follows the transit of the sun and the breathing of the wind.

By 4 p.m. it is dark inside the bothy. Out in the glen, the light lingers, so I head along the semi-frozen burn to catch a last look at the dying day. I relish the onset of darkness in the mountains; it is a special time. As the sun sinks, I watch the hills change their hue, shadows edge across the glen and the mountains sink until only their outlines remain. This is a time of day I would miss if I did not make the effort to sleep in the high hills. It is a secret world only revealed to the few who dwell here as darkness comes.

This is a good time to see wildlife as the creatures of the day shift clock off and the night shift takes over. In the summer, tireless house martins who have spent the long day sweeping through the air in search of insects for their young can finally rest for the night. They have barely put their heads into their cup-shaped mud nests beneath the bothy eaves when black shapes start to hunt the moths and other insects that take flight in the early evening. These bats are every bit as acrobatic as the house martins, moving effortlessly through their sonar world. I rush out at 'bat time' to spend an hour watching the swirling creatures.

Now, in the dark of this winter's night in the empty glen, that summer spectacle seems a long way off. The December skies are empty save for two black-cloaked ravens who take flight from the old rowan tree when they catch sight of me. They live here all year round, surviving on their wits. The pair circle above, keeping their dark eyes on me as I walk about the glen. These birds are intelligent predators who have gained a sinister reputation as a result of their habit of feeding on carrion. Their bad press seems unfair, as eagles will just as happily feed on dead animals but are revered rather than shunned. In Norse mythology, Odin regularly appeared with two ravens, Huginn and Muninn. These birds were thought to rove the world and return to Odin to whisper its secrets in his ear. I'd be surprised if Odin was interested in my comings and goings. It is testimony to the ravens' resourcefulness that they can survive here in December when most other birdlife has long vanished from the skies. I feel as though I am trespassing in their territory. This is their glen and I am an intruder.

As the sun dips behind the hills, the air chills and the grass is touched with hoar frost. The puddles on the track are glazed by a paper-thin veneer of ice that tinkles as it shatters beneath my boots. Half a mile from the bothy, I pick my way between the pools and tussocks in the failing light. There is a movement off to the south. Three deer, silhouetted against the night sky, are making their way towards the small hill above the bothy to settle in the grass for the night. These are not the well-fed animals I watched in Strathconon during the rut. They are wild creatures who may face a lingering death by starvation if the weather is harsh. I know how shy they are. If they see me or pick up my scent on the wind, they will take flight. The estate I am on now is not run for stalking or grouse shooting. No animals are killed here for entertainment. Deer are shot, but only to manage their numbers so that there is a chance for regeneration.

In summer, the stark contrast between these valleys and the overgrazed glens of most of the Highlands is obvious. Here, the grass is long and lush. The deer population is falling, and young trees are creeping out of the watercourses towards open ground like soldiers emerging from trenches, sensing the enemy's retreat. This small part of the Highlands is moving away from the killing culture that has blighted its glens for so long. The landowner, Hugh Raven, has made a conscious decision to reject the model of a traditional sporting estate, and instead is allowing nature to creep back into the landscape. He is part of a minority who feel they have a duty to consider what kind of land they will hand on to the coming generations. Although the numbers of such landowners are growing, it will be a long time before we see widespread change. Right now, it depends on the whim of individual owners.

Such thoughts trouble my mind as I watch the dark moonlit silhouettes of the deer cross the shallow river. I stand motionless as their ghost-like forms slip silently into the night. As I make my way back over the icy river, the glimmering lights in the window welcome me home. The paraffin lamps give the bothy a golden glow, reflected in the sparkling water of the burn. The chimney breathes grey smoke up into the night air, speaking of warmth and comfort. I push open the bothy door and leave the unbroken silence of the darkness behind.

The next day, I wake early. It is still dark as I light the candles and the gas under the kettle. The windowpanes, etched with intricate frost patterns, glisten in the candlelight as the kettle begins to sing. Steaming tea in hand, buried in my duvet jacket, I step outside to stand in the cold and watch the birth of the winter's day. The daylight is creeping into the glen, pushing against the darkness as if it were a solid thing. It is the stillness of this frost-covered landscape that impresses my imagination. Until recently I was a transient visitor to bothies, spending a night or two and passing on. Now that I write in a bothy, I spend up to a week in isolation. My extended time in these remote places has given me a deeper insight into them. I am three miles from the nearest road. There is no footfall at this time of year; no one strolls by. There are hardly any birds and the air is empty of insects. One day in three, I watch the handful of red deer I saw last night tiptoe down from the hills to drink in the burn and then pass on into a world only they know. On a busy day, I catch a glimpse of the ravens as they patrol their kingdom. This December is colder than average; it has dipped to below -5° Celsius over the last few nights. The frigid air has dusted the hills icing-sugar white. Even the burn that normally babbles with incessant enthusiasm is still and chocked with ice.

During the first few days of this trip, I enjoyed the silence and the stillness. It was good to get away from the constant low rumble of urban living, to escape the sound of traffic, the yells of squabbling children and my own addiction to the babble of social media. When I arrived in the glen, I felt myself change a mental gear, settle into the landscape. I wanted to open my ears to the song of this place, to listen to its beating heart. I grew quieter. I became slower. I began to listen to my drowned thoughts. I felt myself breathing as I relaxed into the place and became part of it. Then, as the days passed, I felt an obscure unease. It was as if there was some secret thing here. At first I pushed it away, went for walks, sat and wrote, watched the fire in the evening and dozed; but the feeling of disquiet pursued my days here. It was as though I could hear a distant sound, but couldn't make out what it was. Every day, the sound grew louder.

While the kettle comes to the boil, I go to the burn to get water. I break through the crust of ice and dip my bucket into the cold water beneath. It is a simple act, taking one of life's essential needs direct from nature. As I fill my bucket, I catch sight of the ruined walls of the old black houses that have lain deserted since this glen was cleared of its people in the early 1800s. At this moment, I realise the source of my unease. The silence in this place speaks not of what is here but of what is not. There is no forest. There are no sun-dappled clearings to house deer, lynx and pine martens, no trees to provide food and shelter for birds. The broken walls which once housed people are empty and cold. This landscape is what remains of a holocaust that has stripped it bare and left nature clinging on to the margins. The unmoving land, the empty air and soundless hills are a testimony to what we have done to this landscape. I know now why the stillness and the quiet trouble me: it is because the silence is the sound of a dying land. This reflection saddens me and yet, despite the destruction that has been wrought on this landscape, I know it can recover. There remains a chance to breathe life into this land if we have but the courage and determination to take it.

***

Time is a matter of perspective. For us humans, a lifetime is measured in eighty years or so. Our memories of former time generally stretch as far as our grandparents, with everything beyond relegated to the dim focus of history books and grainy film. Within our own lifetimes, our perception of time changes. I read somewhere that when we are young, time passes by in drips; by the time we are older, it has become a waterfall. In my mid-sixties I feel as if life is a bus just pulling away from the stop as I try to board. Maybe when I'm older I'll arrive at the bus stop and see the bus disappearing in the distance. One day I won't make it to the bus stop. There's a little sunny thought to brighten your day.

One winter's day, Martin and I are walking along the path to Craig bothy in Torridon, looking forward to spending the night there. Craig was once a youth hostel and so the path to it is well made and pretty level for the three or four miles that it follows the coast. In common with all coastal paths, it has the disadvantage that it is exposed to the weather coming in from the sea. Martin and I have been anxiously watching a grey wall of water heading inexorably our way across the rolling waves. It looks as though the dark cloud has spotted us and decided it can't think of anywhere better to dump its load of several tonnes of water than on top of two middle-aged men scurrying along the path weighed down by large rucksacks. We are doing our best to make it to the bothy before the deluge arrives. As we rush along the path with our rucksacks rattling, I spot an unusual line of indentations running along the sandstone. They look like ancient animal tracks.

'Look at that, Martin. Maybe dinosaur footprints.'

Martin sagely shakes his head. 'No, they can't be. The rock here pre-dates the dinosaurs.'

Somewhere in the back of my mind I knew that. My butterfly brain focuses on one fact for a while and then flits on to something else, distracted by a bright colour or something shiny. Martin has an encyclopaedic memory which files things away. Somewhere in his head he has boxes of old bus tickets, scribbled notes of train journeys and faded rail timetables that he can recite with a moment or two's reflection. I had forgotten that his grasp of time not only extends to the adventures of our youth in the 1970s but far beyond that into the geological periods of history.

Martin peers at the marks I spotted on the rock. 'This is some of the oldest rock in Britain. Maybe 600 million years old. Dinosaurs emerged around 240 million years ago.'

I'm looking at the odd-shaped impressions again. 'So they aren't footprints?'

He shakes his head sadly as if talking to a foolish child. 'Can't be.'

I find this vast period of time unfathomable. The landscape appears so solid and immutable, and yet that is only an illusion. It is impossible for me to comprehend such huge shifts in the pattern of time because this landscape exists in a different time zone. It is difficult to believe that this rugged hillside that slopes into the sea below me is a transient thing. We pass through it, like the shadow of a cloud, while the landscape grows and moves on a timescale we cannot comprehend.

I am musing on such deep thoughts when a large raindrop explodes off the top of my bald head. The cold shock brings my mind back from contemplating the mysteries of the universe to far more important matters: we are about to be soaked. I pull up my hood just in time to protect myself from the arrival of several thousand more bullets of precipitation. We are still a good mile from the bothy as the salt wind brings a cloudburst that empties over our heads in monsoon proportions.

'Let's get going.' Martin sets off at full tilt towards Craig with a dejected groan, having lost all interest in geology.

When you read outdoor magazines, you find images of grinning walkers striding along in pristine rain jackets apparently brimming with joy at being out in the pissing rain. Who are these people? I have never met anyone who actually enjoys rain. You endure getting wet. That night in the bothy, while our waterproofs drip from hooks all around us, Martin and I enjoy the perverse pleasure of moaning about our soaking in front of the bothy fire. The pleasure in rain is not being in it but out of it.

***

We think of humans as the only creature that can change the landscape. We can carve through mountains to create motorways, build canals across continents and create our own lakes. But there is another creature who can also alter the landscape: the beaver. The beaver is the only wild mammal to have been reintroduced to these islands. The progress of its return to these shores is of immense importance if other creatures with less public-relations appeal, like the wolf and the lynx, are ever to follow in its footsteps.

On a cold December day, I take the short trip from my home to the impressive modern offices of Scottish Natural Heritage.[1] Their offices sit overlooking Inverness on a wooded hill known locally as Craig Dunain. The building is a modern, functional office block, but it is faced with wooden slats and has an organic feel to it. Once inside, the visitor walks into an airy atrium with clean lines and a polished stone floor. It is odd to think that this modern office, which could be the home of any multinational corporation, is dedicated to preserving and fostering our relationship with wild places. After reception has issued me with the obligatory security tag to ensure that I don't leave plastic explosive in the toilets, they call upstairs.

I want to find out more about how beavers were reintroduced to Scotland, so I have come to talk to Ben Ross, who is Scottish Beaver Project Manager with responsibility for coordinating the reintroduction of these animals into Scotland. Ben is a lean, fit-looking man in his early forties. His enthusiasm for beavers is immediately apparent and he spends the next hour answering

1 Scottish Natural Heritage have since changed their name to NatureScot.

my questions and explaining the history of the release of these fascinating animals.

Beavers became extinct in the British Isles some 400 or 500 years ago when they were hunted to extinction. Today, if you mention beavers to the average person, they will think of Canada or the wilds of Alaska and perhaps won't know that beavers were ever native to these shores. It seems strange that animals of such significance can have totally slipped from our cultural memory. Just like the lynx, there are no traditional folk tales involving beavers.

Sitting in a glass-fronted office, I explain this to Ben. 'It is as though a creature we lived with for thousands of years never existed.'

Ben smiles and gently corrects me. 'Actually, there is evidence of the existence of beavers all over our landscape if you know where to look. The Yorkshire town named Beverley is a corruption of "beaver", and the town's heraldic crest carries a picture of a beaver next to water.'

As we talk, I begin to appreciate Ben's view that the absence of beaver activity has had a profound effect on our landscape.

'Our river systems today run the way they do because there are no beavers to interact with them.' As Ben speaks, his passion for these creatures is obvious. 'Beavers are river engineers, and over the years will have a profound effect on the course of any river they inhabit.'

Ben explains to me that the aftermath of beavers living on a stretch of river might seem to be a scene of devastation, with felled trees appearing to be abandoned and dams flooding land. However, as beavers move through a landscape, they gradually create wetlands behind them where a wide variety of species can flourish. This helps to create biodiversity. Beavers may inhabit a stretch of river for a period of years and then move on when the tree growth cannot sustain them any longer. But they do not abandon the area, instead returning cyclically over the years, slowly expanding the wetland.

I wonder what it would have been like to walk on a summer's day beside a river in the time of William the Conqueror. Even in those times, I am guessing a large part of the medieval forests would have been cleared for firewood and building materials. Despite the changes in the woodlands, Britain's river systems would have been much as they had been for thousands of years. Beaver numbers would already have been in decline, but evidence of their impact on our waterways would have been apparent everywhere – to such an extent that following the bank of a river would probably have been quite difficult, as a walker would have encountered marsh and wetlands along much of its length. I imagine that these habitats would have been home to a myriad of insects, which might have made the riverside an uncomfortable place to be. Beaver-modified rivers would have been ideal for absorbing floodwater and preventing it reaching towns and cities. It's certainly not the case that beavers would solve today's flooding problems single-handed, but they might at least be part of the solution.

Discussions about the reintroduction of the beaver began some twenty-five years ago, but gained only significant impetus in 2005 with the coming together of four organisations to form the Scottish Beaver Trial. The four organisations were the Scottish Wildlife Trust, the Royal Zoological Society of Scotland, Forest Enterprise and Scottish Natural Heritage. After much discussion with landowners, the four organisations chose the area around a series of small lakes in Knapdale.

As Ben explains, 'The area was chosen because it was a relatively enclosed place where beavers would be unlikely to spread from, being close to the sea. The habitat was also right for beavers, with a range of native woodland being available.'

In 2005, twenty animals were trapped in Norway and released into the area. Around the same time, fishermen and farmers began to spot a number of individual beavers in the River Tay. It isn't clear where these latter creatures came from – it's likely they escaped or were possibly released from private collections nearby. Such collections are unregulated, so there can be no control over the level of security, and it's possible some collectors were deliberately casual about how well caged these animals were. Wild beavers living in the Tay was a very different proposition to having them released in Knapdale. While Knapdale is relatively contained, the Tay is the longest river in Scotland, running some 120 miles from its source on the slopes of Ben Lui to reach the sea at the Firth of Tay. It boasts several substantial tributaries and has a catchment area of some 2,000 square miles. In short, once established in the Tay, beavers had room to spread and that is exactly what they did. The presence of beavers in the Tay rapidly became an issue with local people who farmed its banks and fished its waters.

Ben sums up the problem simply. 'If you have the two main landscape-changing animals in the same area, man and beavers, conflict is bound to arise.'

One of the main issues surrounding the catchment area of the Tay is that it contains some of the best farming land in

Scotland. Although it is extremely fertile and valuable, much of this land is low-lying floodplain, which means it is by definition flat and often drained by a series of parallel drains running in to a long, deep ditch at ninety degrees.

'Beavers, of course, are very fond of damming ditches,' Ben explains. 'Even raising the level of the main drainage ditches by a few inches has the potential to flood large areas of land. The wild beavers in Tayside were very quickly having a big impact on the geography of the humans they came in contact with. Concerns began to be raised by farmers and fishermen about the impact wild beavers were having in the area.'

He continues, 'Everywhere in Scotland is to some extent managed habitat. If you introduce a creature like the beaver, it'll have both positive and negative effects on natural and managed habitats. Beavers may create a beautiful wet meadow area, great for insects and a range of other species, but other habitats might take a hit. For instance, aspen is a rare and important tree in Scotland, but beavers can't get enough of it and that can impact local populations. Or it might be that there are rare lichens growing on trees in an area, and if beavers turn up you can lose all that.'

These are the kind of issues that faced the rewilding programme for beavers. The relationship between beavers and the landscape is complex. Beavers may well chew down many of the trees near a watercourse.

Ben also outlines the impact of deer numbers on the habitat. 'Native trees have adapted to cope with beavers. They'll sprout new growth from a chewed stump, effectively coppicing and creating a bushy woodland. But they can't do that if there are too many deer.'

I often walk in the woods beside the River Ness. As Ben is speaking, I begin to realise that these woods are doomed. They are populated by tall, mature trees, but the forest floor is bare, the result of overgrazing by the roe and red deer. I'd never thought of it in that way before. There are no young trees coming up because the deer cut them down. I am beginning to see how this great jigsaw of life fits together and to realise just how complex and interdependent everything is. The damage done by our out-of-control deer population is huge.

I put to Ben a question that has been puzzling me for a long time and has proved controversial amongst those interested in beaver rewilding. 'I read that on the very day that the Scottish parliament agreed to allow the beavers of Tayside to remain, Scottish Natural Heritage issued nineteen licences to allow beavers to be shot in certain areas. Surely there's a contradiction there. We've just agreed to allow a native species to return to these isles and then we agree that they can be destroyed – how can that make sense?'

Ben has a thoughtful approach to the question. 'There's no doubt that beavers can have a critical impact on farmland. We have to decide what we want to do with the land. We have to grow food, and we also want nice woodlands and wetlands for beavers, but we can't have all of this together.'

I'm puzzled. 'But what the hell is SNH doing, issuing licences to shoot a protected species? It doesn't make sense.'

'You have to remember that the beavers that were released in Knapdale were released after a lot of consultation with local farmers and landowners. Those that were released or escaped into the Tay did so with no consultation. The farms in the area just started seeing beavers and having to cope with the effect they had on the landscape. The floodplains that are affected are amongst the most productive in Scotland. We are putting in and trialling measures to minimise these impacts, and they work well in some cases but not in others. We have to grow food somewhere, so where there is no other solution, licences were issued to shoot beavers only in areas of prime agricultural land.'

In 2015, the Scottish government had to make a choice. Beavers were free in the landscape and would inevitably spread. There were three options: to eradicate them completely, to let them spread naturally or to actively assist their spread by relocating them across Scotland. Roseanna Cunningham, the environment secretary, decided on a policy that would allow the established beaver population to spread naturally across Scotland. At that point, beavers were not a protected species, and it took another three years before they were given that status. In that period, there was an enormous amount of negotiation with interested parties such as farmers and landowners. It's very rare for conservation bodies and landed interests to come together – they are normally poles apart – but in this case they did. The National Farmers' Union, Scottish Land & Estates, the Scottish Wildlife Trust and the Royal Zoological Society Scotland wrote a joint letter to the Scottish government accepting that beavers should be a protected species with the proviso that the legislation included a pragmatic system for the granting of licences to cull beavers where they became a problem.

To go by press reports, the farming community seems to be against anything that seeks to protect or promote the environment. If you believe the accounts you read in the media, we have a farming system that can only flourish if any kind of wildlife is eradicated. Landowners and farmers fought against the proposal in the Scottish parliament to introduce controls on the spread of hill tracks, and they reacted in horror to the reintroduction of white-tailed eagles in England and still campaign against them in Scotland. Listening to Ben, I begin to see that my view of the farming community as the enemy of rewilding is simplistic. He is convinced that the key to the successful reintroduction of species is an effective conflict-resolution process that can deal with the issues that arise when newly introduced species inevitably collide with the interests of their human neighbours.

It is clear that there is a delicate balance to be found between our desire to see beavers flourish in Scotland and our need to keep our agricultural land productive. The outrage felt by some in the conservation community about the legal culling of beavers is understandable, but the alternative – moving beavers from an area where they are coming into conflict with agricultural interests to another, less sensitive area – is not without its challenges. So far, the reintroduction of the beaver has been notable in bringing the disparate interests of the farming community and conservation bodies into relative accord. Moving beavers from one area to another would require bringing the landowning interests of new areas together, which is rarely an easy process. The debate between lethal control and translocation continues, and more research is needed to examine the effect of beaver culling on the population as a whole and to examine the benefits of moving them elsewhere as an alternative.

It's clear that for the first mammal to be reintroduced to the British Isles for over 400 years the future is bright. Beavers are here to stay. I wonder how long it will be before I can take a walk down the River Ness and find a beavers' lodge upstream. I hope that one day the dams that beavers construct will be a common sight on our rivers, and the wetlands that they create will provide habitats for a wide variety of creatures. It's a pity that we have no children's bedtime stories about Mr and Mrs Beaver struggling to keep their dam watertight against the rising flood waters. Perhaps I should write one.

***

December stays cold, the weather pattern fixed by a high-pressure system parked over Iceland. Early snow has turned the summits of the hills white. The chance of a taste of winter has me studying the forecasts intensely and pulling my cold weather gear out of cupboards. Late December is not the best time to wander the hills of Scotland. The days are frustratingly short and the glittering snow cover has not had time to thaw and refreeze, so it will be soft and brutally slow to walk through. I know all of this, but the lure of the cold is too great. I take my winter boots down from the shelf where they have languished all summer. They feel heavy and solid, and there is something reassuring in that. I always look forward to putting my big boots on at the end of the year. It's a sign that winter has arrived.

The crust on the fresh snow holds my weight for a tantalising moment before it fractures and I drop with a crunch into the slush beneath. With every step, my pack of coal, food and equipment grows heavier. I had forgotten what it feels like to carry such a load through deep snow; now the ache in my legs is reminding me. I am on the second day of a trip into the heart of the Cairngorms. The scale of this place dwarfs everywhere else in Britain. Broad, sweeping glens reach out to touch the feet of huge rolling hills that rise into vast open skies. In winter, it is an empty place. If you wander toward the centre of the Cairngorms on a winter's day, you leave our cluttered world of tarmac and concrete behind. You enter a place of silent corries and savage, unforgiving winds. The weather here is the harshest in this island. In these cold months, this place demands respect and can exact revenge on those who treat it with disdain.

'Ach – here, man, let me break the trail,' Neil calls from behind.

Neil is a wiry Aberdonian in his early fifties who has spent more time in these hills than he cares to remember. He has volunteered for the Mountain Bothies Association for many years and is a Trustee.

He turns as he wades past me through the snow, his grey beard jewelled with frost. 'Ah'm nae carrying what you've got in that bloody sack.'

Tactfully, he only mentions my load, and not the fact that I am also a good ten years older than him. He only has a day sack compared to the behemoth I'm carrying, but it's a generous gesture nonetheless. It's obvious that there's a warm heart beneath that dour exterior.

'It's not too bad,' I lie. I'm grateful as he forces his way through the snow, making it easier for me to follow.

Neil plods on in silence. We walk for another twenty minutes. As we climb, the temperature drops noticeably. We have crossed an invisible line – now the terrain beneath the snow is frozen and the sound of trickling water in the burns we pass is silenced. Despite Neil taking the lead, I realise that I am slowing down.

I run over my planned route in my head. I'm almost at the Hutchison hut, one of the highest and most remote of all the MBA's bothies, but that's the easy part. It will be followed by a day-long walk to the next bothy, the ultra-remote Faindouran. Two years ago in winter, it took me two attempts to get to that lonely shelter. On my first try, I was stopped by deep snow and forced to spend the night in the emergency shelter at the Fords of Avon, the hub of the great glens of the Cairngorms. From Faindouran, I intend to head north and emerge from the hills at Glenmore Lodge Outdoor Centre. Another long and exposed day. From there, I can get a lift to Aviemore and the train home. By the last day, I will be travelling light, but the route climbs over the exposed shoulder of Bynack More at 2,600 feet – that's not a place to be late in the day in the teeth of a Cairngorm blizzard. Mountaineering plans are made in a warm room with a map spread out in front of you and a glass in your hand. Mountaineering decisions are made on aching legs, sweating, gasping for breath with reality howling about your ears.

I tighten the hood of my cagoule around my face, but still the icy wind numbs my cheeks. The last few hundred metres to the bothy are a struggle, fighting our way through deep snow-drifts. The Hutchison memorial hut sits alone in this icy corrie and is as lonely a place as it is possible to imagine. Coated in ice, the single-room bothy looks like a polar explorer's shelter. It is surrounded by windblown snow with ice-glazed cliffs rising behind. I have to kick the drifts away from the door so Neil and I can enter. Inside this basic refuge, we are out of the wind, but the temperature struggles to rise above freezing. There is a bench running along one wall, a low metal table and a serious-looking solid fuel stove in one corner.

I cast my eyes around the austere shelter. 'It's smaller than I remember. You couldn't get more than four folk in here for the night.'

Neil laughs. 'Ach, no. You can get nine in comfortably. It jist takes some planning.'

He goes on to demonstrate. 'Two under the benches, two on top and five head-to-tail on the floor.'

I'm not convinced. 'Really?'

Neil sits down and begins to take out his lunch. 'Aye, there was sixteen in here one night a few years ago.'

'Bloody hell! What was that – the dwarf mountaineering club?'

As Neil and I sit silently eating our lunch, I begin to assess the situation. I have learnt, through my time in the Cairngorm Mountain Rescue Team and my own experience in these mountains, that you have to respect the Cairngorms. If you take them lightly, they can kill you. It's taken longer to get to the Hutchison hut than I expected. The conditions are bad. It's early winter and the snow is soft and unconsolidated. If it snows again tonight – and from what I can see from the hut window, it will – walking through these hills could become a tortuous business. If I stay the night here and continue to Faindouran as planned, a heavy snowfall could trap me. I'll be alone in a remote place, totally reliant on my abilities to get me out. The slight tickle I felt earlier in my throat is becoming more persistent and I am beginning to shiver now and again.

I'm mulling this over while I eat my sandwiches from the low table, when I notice Neil looking uncomfortable.

I put my sandwich down. 'What is it?'

Neil squirms a little. 'Ah, nothing really. It's just maybe dinnae put your sandwiches down on that table.'

I examine the table. 'Why? It's a fine piece of wood.'

Neil nods. 'Oh aye, it's a fine piece of wood, ye ken.'

'Then why not use it?' I ask.

Neil is silent for a moment and then speaks hesitantly. 'I was here when this bothy was renovated.'

'I see. And … ?'

'Well, we set up a temporary latrine over yonder. Not much more than a wooden seat with an oval hole in it, screened for privacy.'

I am beginning to feel uneasy, wondering where this is going. 'All those folk working up here, I suppose you'd need one.'

Neil coughs and scratches his greying beard for a moment. 'We did. I used it quite a few times myself.'

I sense he is holding something back. 'I don't see what this has to do with the table.'

Neil has the look of a man in a confessional. 'A few months after we finished the bothy and tidied everything away, I came back here and was having my lunch. I took a look at the table and something seemed familiar. So I looked at it a little more closely and then turned it over … '

He turns the table over so that I can see its underside. There, cut into the base of the tabletop, is a large oval hole. It can only be one thing: a toilet seat.

I feel my stomach roil. 'Ah, I see.'

I pick up the remains of my sandwich, wrap it carefully and put it in my rucksack. For some reason, my appetite has deserted me.

We both sit in silence surveying the table until I speak. 'Well, it is a fine piece of wood.'

Neil nods sagely. 'Pity to waste it.'

I glance out of the bothy window and notice that the weather is deteriorating.

'You know, Neil, I'm having second thoughts about this,' I say, finally giving voice to my concerns.

Neil looks at me carefully, knowing that this decision could be critical. 'Aye? Well, it's a long way if this weather turns.'

'Yes, it is. And there's not much daylight and I'm not sure I am feeling 100 per cent.'

Neil presses his face to the frosted glass and stares at the sky, which is rapidly filling with snow clouds. 'Aye, it's nae looking so good.'

He has been coming to these hills since he was a boy and has almost fifty years of experience here. He too knows that the Cairngorms demand respect.

I cough again, the slight tickle in the back of my throat growing more insistent. 'That snow'll make slow progress.'

Neil nods. 'Not consolidated yet. Maybe it'll be easier later in the year.'

I nod, and that is the decision made.

On the long walk back down the glen, Neil begins to talk about the Cairngorms he loves. He points out things that I would never have noticed.

'You see those small ridges of earth down by the Derry Burn?'

By the river are a pair of roof-like ridges, picked out by the new snow. 'I can see them. Either side of the river?'

'There was a dam there once. They built it to float timber down the Derry Burn.'

Hearing him talk, his intimate relationship with the area is obvious. This is never more apparent than when he speaks of the origins of Bob Scott's bothy. In these empty places, men who have spent their lives here leave a mark. They become part of the folklore of these mountains in ways that are impossible in teeming cities. As we walk back to Bob Scott's, Neil begins to tell me about his memories of the southern Cairngorms. He takes me back to a time in the early 1970s, a time when I was beginning my own journey in exploring the hills of the Lake District and he was finding his feet in this wild place.

'In those days, we'd camp beside the river at the old Canadian loggers' camp, ye ken. It were jist an area of flat grass the Canadians had abandoned. We'd spend the evenings in the bar at the back of Mar Lodge, which is nae far from the campsite.'

Neil describes this as a haven for climbers and walkers in the 1960s and 1970s. It was a long, narrow room with a bar at one end and a free-standing fireplace at its heart. The bar was a gathering place for all the folk who lived, worked and played in the hills that meet at the Linn o' Dee. Thronged together in the warmth of that smoke-filled room were climbers in their ragged pullovers and army surplus boots. Next to them, clad in their uniform tweed, were the ghillies, gamekeepers and stalkers who worked on the surrounding estates. Donkey-jacketed forestry workers had their own huddle in one corner.

Neil remembers one man at the centre of this community

of hill folk. 'He'd his own chair that nobody else sat in, the King of Derry, Bob Scott.'

Scott was the gamekeeper for the Mar estate and lived in a remote cottage called Luibeg. His cottage was a few miles up a track where a large burn of the same name met the main force of the Derry Burn at Derry Lodge. Bob was a sociable character who was always happy to welcome anyone who wandered by and supplemented his wages by running a small bothy for walkers.

By the 1970s, Bob had become a Cairngorm legend, and nights at his bothy were shared by famous outdoor folk like Tom Weir and a young Tom Patey. The late Adam Watson knew Bob Scott well. In his fascinating book on the Cairngorm mountains and the history of the folk that climbed them, *It's a Fine Day for the Hill*, Watson relates a number of tales that give an insight into Bob's character. Bob was fiercely protective of his beloved Cairngorms, possessing a fierce turn of phrase which left anyone who abused them in no doubt as to his opinion. Watson relates that on one occasion Bob spotted someone taking an egg from the nest of an oystercatcher. Bob called out and the man quickly replaced the egg. Bob raced over to him and made his feelings plain. 'Haud doon that glen, ye bugger. Ye'd better nae come back or I'll string ye on a tree for the heedie craas [hooded crows] to pick. Tell yer bloody freens [friends] nae to hairy [harry] oor birds or they'll get worse.'

For a while Derry Lodge, about half a mile from Bob's cottage, was rented by the Cairngorm Club and used as a mountain hut. The lodge is now derelict, but was a large and substantial building then. Bob was passing in a blizzard with the weather deteriorating fast. He called in to the lodge and met two teenage boys who were planning to head down the glen that night. Bob advised them that it would be dangerous to walk out in that weather and they should stay put until the following morning. That night the wind grew ferocious and large drifts built up around his cottage. Bob was dozing by the fireside when he was awoken by a knock at his door. The two teenagers staggered in from the storm saying that the nights they had booked at the lodge had expired and they'd been put out by the warden. Bob brought them in, sat them beside his fire, and thawed them out before giving them a bed for the night.

The following morning, Bob marched through the snow and confronted the custodian of Derry Lodge. 'Fit wye did ye pit oot the twa lads? God almichty, it was nae a nicht for a beast to be oot in, let aleen twa young loons. Fit if there had been nae bothy for them? Ye'd have landed in a richt mess if ye'd forced them oot and they'd got smored to death in the snaa on their wye doon the glen. Ye wouldnae put your ain wean oot on a nicht like that. Ye should be black burnin' ashamed o' yersel'!'

Neil and I have been battering our way through the darkness and snow for several hours. At last, as we enter the forest, a faint light shows through the trees. It is the bothy built in Bob Scott's honour. I am glad to push open the door and sit down in warmth. Neil and I stayed here last night, and the other three men who were here then have spent today walking out to the village of Braemar to restock the bothy's coal. My legs are more than a little weary when I sit down beside the fire and someone thrusts a cup of tea into my hands. The previous bothy burnt down and was rebuilt in 2005. When it was reconstructed, by volunteers funded from public subscription, the walls were insulated with sheep's wool. No matter how cold it is outside, the enormous pullover in the walls keeps the occupants snug. We spend the night swapping tales and drinking whisky in the warm glow of the fire. As I sit in comfort, a little part of me regrets that I am not sitting in splendid isolation high in the remote Hutchison hut. It is only a very small part.

The next morning, the irritating tickle in my throat that I was trying to ignore yesterday has turned into a full-blown cough. Neil is croaking too. It snowed last night and the ground beside the bothy is frozen hard. Low in the glen, there is a light breeze and the day feels pleasant, but when I look high on the hill I notice bands of fleecy snow clouds racing across the broad Cairngorm summits. They are enough for me to know that carrying on in the poor conditions could have landed me in a serious situation, especially as I'm now struggling with a cold. Sometimes the best judgement in the mountains is to retreat and wait for better weather. As we walk back down towards the Linn o' Dee, both spluttering and coughing, I have the feeling I may have dodged a bullet.

When we meet the road, we take a short detour towards Mar Lodge and the stone bridge that straddles the Derry Burn just before it enters the River Dee to eventually flow into the sea at the city of Aberdeen. Burn is the wrong term for this watercourse now, as the water roaring before us, tumbling over the rocks, is a full river. On the bend in the river below, Neil points out where the old Canadian logging camp used to sit. When he was a boy, it was a clear area of ground, but now it has been swallowed by the forest and hosts mature trees. It's hard to believe that it was once an established campsite where the folk of Aberdeen would escape into the hills. Neil has a bond with this place. Over forty years after his childhood trips, he still returns to these hills and delights in his wanderings. The boy who played here in the heady days of his youth has not left; he stands beside me now. In the Aberdonian dialect, he is the Cairngorm mannie. The Derry Burn runs through his veins.

At Perth station, the Inverness train rolls in. It's on time but crowded, and I have to squeeze myself into the carriage and take the only free seat. I am still dressed for the Cairngorm winter, so I am soon sweating on the warm train. We've only been travelling a little while when I begin to get odd looks from my fellow passengers. I get the distinct impression that they are surreptitiously sniffing the air. When I press my hand to my rucksack I feel a damp patch. I pretend to scratch my face and when I sniff my hand there is the unmistakable rich aroma of whisky. The bottle must have leaked in my bag. The other passengers shuffle uncomfortably. I realise I must smell of the bothy. I have that unique bothy blend of sweat and smoke. The other passengers are looking at an unshaven old man, dressed in ancient outdoor clothes, reeking of smoke and drink. An alcoholic tramp, probably, who's somehow got on to the train by avoiding the ticket barriers. I take a secret pride in knowing that I am slowly filling the train with the stink of the wild.

For the next two weeks, over Christmas and New Year, I am consumed by a bad cold and unable to walk further than the shops. My visit to Bob Scott's bothy and the glen he loved has made a deep impression on me. Bob Scott has left an indelible mark on the area – perhaps Neil will too. I wonder about my own legacy. Will I leave the hills better or worse when I have finally hung up my boots?

My relationship with the hills and glens of my Highland home is different now, deeper than it once was. I understand the landscape far better than I used to, and such understanding has brought a different perspective. Where once I saw wilderness and stood in awe of a magnificent landscape, now I see desolation. Now I see the scars we have wrought on this earth. I see the high moorland burnt for the shooting of driven grouse. I see our hills denuded of trees. I stand in empty glens where birdsong is just a memory. I cannot unsee these things. I can't push them away into the back of my mind and return to the naivety of my youth; I can only stand and weep for this land.

The time has come for me to do something for this landscape I love, to stand up and add my voice to those calling for reform, for an end to treating this place as one great sporting estate where rich men play out their killing games. Now we must put back what we have taken from the land, and give nature a helping hand rather than batter it with the shotgun and the bulldozer. The time for change has come.

JANUARY

I'm standing in the echoing foyer of Inverness station, surrounded by crowds of young backpackers and tired-looking businessmen. The Edinburgh train slowly pulls into view. The doors open and the platform is quickly crowded with a multitude of travellers. The afternoon train is busy with groups of women returning from shopping trips to Scotland's capital and commuters coming home from their Friday meetings. Folk jostle on the platform, and if I was meeting someone else it would be easy to miss them. Martin is on this train and I'm confident he'll stand out in the throng. Even though I can't yet see him, I know that somewhere in the crowd he is struggling with his enormous rucksack and probably scattering fellow passengers as he does so. Once he shoulders his rucksack, it will be impossible to miss him. He is six foot three and his rucksack towers above him, topped, as always, with a roll of Karrimat balanced horizontally. No backpackers today would walk around with their sleeping mat rolled high above their pack, but that is what hikers used to do in the 1970s and Martin steadfastly refuses to leave that decade. He is a great believer in the adage that you shouldn't fix something that isn't broken, so if it worked in 1974 why change it now? If that didn't make him tall enough, his giant sixty-five-centimetre ice axe, strapped vertically to the back of his rucksack, will make Martin and his pack a seven-foot giant.

After a few moments, I spot a rolled-up Karrimat with a metal spike protruding above it making its way towards the barrier, floating above the heads of the other departing passengers. Martin is making his way through the crowd beneath the rucksack of doom. There is a rail official standing watching people coming through the automated barrier. Why someone has to watch over the machine that does their job I have no idea. Surely if it takes a person to watch the automated barrier, it would have been simpler to let the human collect tickets and not bother with the barrier. I decide to file that particular question under 'things I'll never understand'. I have sympathy for the middle-aged man overseeing the barrier, his eyes glazed with boredom. I know something he doesn't: he is about to have a confusing conversation with an awkward customer.

Martin arrives at the barrier with an irritated expression on his aquiline features. He reaches into the pocket of his tartan hill shirt, produces his ticket and waves it at the rail worker.

The rail worker wakes from his dream and helpfully indicates the little slot where the ticket goes. 'Just pop it in there.'

'I want to keep it,' Martin replies, his tone of voice suggesting he is prepared to die for the small piece of cardboard.

The railwayman looks confused and resorts to repetition. 'You have to put it in the machine.'

There were no automated barriers in 1974, so Martin won't use them.

'Souvenir – I want to keep it,' he explains to the puzzled man and proffers the ticket for examination by a human being, which is the way it would have been done over forty years ago.

'It's regulations … ' the perplexed official begins, as a tidal wave of humanity builds up behind Martin's rucksack. Then he realises that he is up against an irresistible force and does the only logical thing: gives up. He presses a magic button, the immovable barriers open and Martin escapes into the world beyond, still clutching his precious ticket.

Martin 2, Rail authorities 0.

Once safely ensconced in my riverside flat, lethal ice axe stashed in the cupboard and giant rucksack parked, we get down to the matter in hand. Martin sits pondering the map stretched out on my dining room table. One of my great joys is allowing my imagination to run free across the landscape of a map. There is a world before me in these contours. There are hills, valleys and rivers, lochs, fords and bridges, and I can explore them all in the comfort of my home.

'Have you been on the hills much since Christmas?' he asks casually, refilling his glass with red wine.

I shake my head sadly. 'No. January's been a disaster. Easterly winds all the bloody time and mild.'

Martin nods sagely. 'Been the same in the Lakes; I've been nowhere since I was here in November.'

'And I've had that damn cough I picked up in the Cairngorms since before Christmas. I've only just got rid of it,' I tell Martin as I unfold the map of the Monadhliath mountains.

He leans over and begins to study the contours. 'There's some nasty bugs around these days. Have you heard about that virus thing in China?'

I can vaguely recall something on the news, but I didn't take much notice at the time. 'Yes, I heard about that. It's just a scare story. It'll come to nothing.'

Martin nods in agreement. 'I'm sure you're right.'

'You don't normally come up here in January. Why this time?'

Martin smiles. 'Make hay while the sun shines. I'm making the most of it while I still can.'

I'm surprised by his fatalism. 'You've years yet.'

He rubs his sore knee. 'I'm not so sure about that. Where's this bothy, then?'

I point to a place a few kilometres from where James and I did our mountain hare spotting just south of the River Findhorn.

Martin examines the area at the end of my fingertip. 'There's nothing there.'

'There's nothing on the map.' I wave my map finger in the air to emphasise the point and open my laptop. Google Earth opens up and I find the same spot we were looking at on the paper version. 'Look there.'

On the computer screen, just visible, is the outline of something square that might be a building seen from above.

Despite Martin's distrust of technology, even he uses the imagery offered by Google to support what information he can find on the Ordnance Survey map. 'Hmm, that could be anything. A ruin, maybe.'

As an avid bothy hunter, I have learnt how to interpret the images I see on Google Earth. 'I don't think so. Look, the lines are sharp. Ruins are not usually that clear.'

Martin is unconvinced. 'How do you know it's got a roof?'

Now we are getting into the realm of the mystical. 'Here's a trick: look at the shadow the thing casts. It comes to a point. That's a gable end. It has a roof.'

It is to Martin's credit that he is willing to trust me on such expeditions. He understands my passion for finding hidden bothies and, even if he doesn't share it, is prepared to indulge me. I'm not sure how this little passion of mine began, but it keeps me poring over maps for hours, looking for off-grid mountain shelters. I suppose it started when I discovered that there are many more bothies in Scotland than the 'official' ones maintained by the Mountain Bothies Association. There are a few hundred usable shelters over and above those the MBA officially curates. Some are crude, draughty shelters, while others are solid and comfortable. Some are open secrets whose whereabouts are widely known; others are only spoken about in hushed tones. It's comforting to know that even in this internet age when information rains out of the ether and all that can be known is available at the click of a mouse, there are places where folk can sit beside a fire and close the door on Google. It will be a sad day when the last secret bothy is gone.

Martin looks at the object on the computer screen and then back at the Ordnance Survey maps. He trusts paper, but is sceptical about its digital equivalent. 'Even if it is there, it might be locked.'

I nod. 'Well, yes. In the end, if you spot one of these things, you just have to go there; there's no way round that.' I hesitate, wondering if Martin will back out.

'Oh, all right, then,' he says, almost managing to sound enthusiastic. 'Let's go and take a look tomorrow.'

I settle back in my chair, content that we have a plan of action. 'More wine, Martin?'

***

The following morning, after Martin has repacked his collection of plastic bags and I have deposited around seven kilograms of fine house coal in my rucksack, we head south down the A9, the main artery of the Highlands. In the distance looms the Cairngorm massif, enveloped in great grey monoliths of cloud.

Martin examines the view sceptically. 'Weather doesn't look very promising, does it?'

'Nah, it's been a lousy winter so far. Here, check the forecast.' I pass him my phone.

He takes it from me and examines it as if it's a piece of technology from Mars. 'I can't use this.'

I had forgotten they didn't have mobile phones in 1974. Martin, in common with around two thirds of my generation of baby boomers, refuses to have anything to do with the technology of today. Many of my friends shun the advance of technology as if it were infected with some terrible curse. New technology set me free to bombard the world with my outpourings, but for many of the folk I know it's a dark mystery that they don't want to spend time trying to understand. When your curiosity dies, you die with it.

I laugh with him. 'Well, we'll just have to find out the hard way, won't we? I've decided I'm going to make something of this winter, no matter what happens. So far, most things have gone wrong. I've failed to see whales, beavers and pine martens. I've had a lousy cold for a month and it's been windy and mild, but I am determined to make something of this season.'

Martin hands me back my phone with a sigh. 'At least you've got the rest of the winter ahead of you.'

I nod. 'Yeah, what else could happen?'

We turn down the narrow single-track road that follows the River Findhorn into the heart of the Monadhliath mountains.

When I visited these hills in November with James, they were in the grip of an early winter freeze. The River Findhorn was choked with ice, its passage through the landscape stilled by the cold. Today the ice has melted and the air has a warm, moist feel. Now the river that ambles through this broad glen, between these rounded hills, babbles unimpeded over the rocks. A few patches of snow remain clinging to the summits of the mountains, a fading memory of colder days. Above us, ranks of grey clouds sweep across dark hills and the low light flattens the features of the landscape.

Martin shoulders his massive rucksack. 'I hope this won't be like last time.'

At first I wonder what he means, and then I remember. The last time I brought him to a high bothy in these hills, we found the place almost buried in a monster snowdrift. Tonnes of snow had been driven across the tops of the hills and dumped on top of the bothy, which sits in a small depression. Only the door lintel remained visible above the snow. Since neither of us carried shovels and the quantity of snow was too great even to be shifted by the mighty Excalibur, we had to head home.

I laugh. 'No, there won't be that amount of snow, I am sure of that. I've never seen a bothy buried like that before or since.'

The hills that were white in November have turned brown with the receding snows, and this is bad news for the mountain hare. Now their white coats make them vulnerable to the talons of golden eagles that sweep out of the skies like avenging angels.

My pack feels unusually heavy with its stocks of coal and food.

It's a mild day for December, but I hope to be sleeping high and it will still be a chill night. I am carrying the luxury of a duvet jacket, the bothygoer's secret weapon. Nothing quite beats the comfort of a down jacket. Warm even in the chilliest night, they can bring comfort to what would otherwise be hours spent shivering. As usual, I find myself cursing the extra weight, although if I am able to get in to the bothy for the night I'll be grateful of its warmth.

Before I leave the car, I put a note in the window:

CAMPING IN THE HILLS
BACK TOMORROW 22.01.20

Martin looks puzzled. 'Why a note? We're not camping.'

'People get curious and they might think someone hasn't come back to their vehicle. Maybe got lost. Might call the rescue on me.'

Martin looks unconvinced. 'Okay, but *camping*?'

'Some of these estates don't welcome visitors. I don't fancy four ghillies turning up in a Land Rover at the bothy in the middle of the night. So if I say camping, then it sort of puts them off the scent.' I can't help scanning the road for shiny green vehicles full of tweed-clad, shotgun-carrying estate workers.

There are rumours that some of the keepers in these hills are less than scrupulous about the laws protecting birds of prey. Hen harriers and golden eagles have vanished off the radar in this area. For that reason, visitors are often treated with suspicion and even hostility. I don't want them to know that I am bothy-hunting.

Martin looks a little concerned, although considering his ability to face down half the staff of the rail network with ease, he needn't be. 'Is that likely to happen?'

I laugh. 'No. Anyway, I am camping in the hills. Just inside a bothy.'

We pass over the bridge where James and I saw the barn owl in November. Sadly, there is no repeat performance today. The path rises slowly up the side of the hill and over a low saddle. I'm surprised by how easy the climb is. The cold that had clogged my lungs for the last few weeks has finally departed. Over the brow of the hill, I spot half a dozen mountain hares, their white coats bright against the dark peat hags. When I was here with James, they were all but invisible to the naked eye, but now the snow has melted I can see them with ease. I wonder if the hare I got so close to is watching me now. Hares tend to stay in the same place, James told me, so it's highly likely. Today is not a day for stalking hares, though; we're here on a different mission.

I point out the mountain hares to Martin and he takes out his camera. 'Oh damn, I'm out of film.'

I watch Martin as he fiddles with the intricate workings of his camera. 'Can you still get film?'

Martin raises the camera and captures an image of a hare who has obligingly remained still. 'Oh yes, if you know where to go.'

I imagine him tracking down the last few rolls of film at the back of a storeroom in some ancient shop in the lost back-streets of Bolton.

By the time we reach the brow of the hill, we have a thousand feet of climbing behind us. On the exposed flank of the hill, the wind catches us, peppering us with raindrops as it sweeps in across the dark green of the glen below. This is the open plateau of the Monadhliath and there is nothing to stop the wind coming twenty-odd miles across this empty place.

The coal in my rucksack is growing heavier with each step. Now we are above the snow line, and here and there patches of hard, old snow cover the track. Both of us fall silent now, locked into the effort of the climb. The wind was mild in the glen, but now it chills us. I pull my hood around my ears as the mist begins to thicken and obscure the landscape. My sticks and boots hit the ground in a regular rhythm. There is no secret formula to reach a high bothy – it's just a steady effort, a relentless push against the rucksack straps. I'm breathing hard as I crest the ridge on protesting legs. I lean on my poles, catching my breath, glad to see that the ground is falling away in front of me. The fine mist of rain obscures everything, condenses on my jacket and jewels the heather.

I turn to Martin a few metres behind me. 'We're almost there. It should be somewhere here.'

Martin stares dubiously into the void. 'If it's here at all, that is.'

He has a point – I can see nothing but the grey misty rain before us. Perhaps I am wrong. Maybe the image I saw on Google Earth was more in my imagination than in reality. I am beginning to despair when I think I spot something angular on the track below. Moments later, whatever it might be is swallowed by the mist. I don't say anything to Martin in case I am mistaken. Two or three paces further and the mist clears to reveal the roof of a bothy.

'It's here!' I cry with a childlike surge of enthusiasm.

A black chimney stands above the roof. At least I won't have wasted all that effort carrying coal up the hill. The wooden bothy looks reasonably new, less than ten years old, perhaps. It has a spacious veranda which overlooks the bend of the river below. It has been well constructed and I'm sure we'll be comfortable inside.

Martin is impressed. 'This is quite a place.'

We are both elated. There is only one question: can we get in? I turn the handle of the big wooden door. It doesn't move. My heart sinks. I try again, but it is locked solid.

'Oh bugger.' Martin is despondent.

Now the light is fading and it has begun to rain. Peering through the windows, I can see a big stove, tables and chairs. Desperate to save our bothy night, I commence a methodical search for a key. I turn over stones, search in the gutters and tap every plank I can find. After twenty minutes, I am forced to conclude that if there is a key, I can't find it.

Martin sits patiently on one of the wooden chairs on the veranda, watching my efforts. 'It's no use, we'll have to go down.'

I search stubbornly for a few more minutes, but eventually I have to concede. 'Aye, all right. There's no key. Still, it's all part of the game.' I try to sound cheerful, but inside I'm miserable.

I dump the coal outside the locked bothy door – a little present for the estate workers from the coal fairy – and we head back down the hill. I'm taunted by the fact that as I trudge back through the darkness, rain seeping in around my neck, I could be sitting in the bothy watching the first of the yellow flames licking about the coal and looking forward to a warm, dry night with my old friend. By the time we are halfway down, darkness has descended and we are following our head torch beams. Suddenly, a white shape hurtles towards me. It is a mountain hare, caught in my torch beam and disorientated. We almost collide and then he heads off into the night.

'At least we will get back to Inverness in time to have a meal and dry out. Shame we couldn't get in. Nice bothy, that.'

Martin senses my despondency. He too has wasted his day, but he tries to console me. 'Ah well, there's always another day.'

Indeed there is. A plan is forming in my mind.

***

The following morning I call an estate office number, but there's no answer. I have a bothy in mind, and as this is Martin's last day I don't want to risk another fruitless walk. I don't think even his patience with my bothy obsession can take that. I'll be able to tell a lot by the reception I get from the office. This is one of my secret bothy-hunting techniques. You never know how an estate will respond to enquiries about the availability of their bothies for two old men with overly large rucksacks and a passion for fireside chats in remote places. Some are incredibly welcoming, and even offer to pop up and supply the place with logs; others give the distinct impression they'd like to murder you.

Martin is in the kitchen frying a breakfast packed with enough cholesterol to kill a regiment. A huge fried breakfast has become a tradition when he comes up for a long weekend, and over the years Martin has perfected the art of the fry-up.

I dial the estate office again. This time a woman answers, her voice redolent with a Scottish public-school accent. 'Hello.'

I take a deep breath. 'I'm thinking of spending a night in the Black bothy.'

Silence.

I take another breath and detect the scent of tweed and wet Labrador coming through the receiver. 'Is it open?'

There is a muffled intake of breath. 'The Black bothy?' I know what she's thinking: why the hell would anyone want to spend a night in there?

I'm determined not to be put off. 'Yes, the Black bothy.'

After a pause. 'It's not that sort of bothy.' *This is some lunatic who wants to spend a night on our estate for free.*

As far as I am concerned, all bothies are *that sort of bothy*. If they aren't locked, you can stay in them. That's the rule. 'Is it locked?'

Now he wants to know if it's locked. Don't tell him it's open, for God's sake. Put him off. She hesitates just a fraction too long before answering. 'Och aye, that's right. It's locked.'

I don't hesitate. 'Oh well, if it's locked I'll not bother, then.'

She relaxes on the other end of the phone. 'It's locked, yes. That's it.' *Go away – we don't want your sort round here.*

'Oh well, thanks anyway.' I hang up.

It's open; we're going in. It's a widely accepted convention amongst the bothy-going community and most Scottish estates that if a remote shelter is unlocked, you can sleep in it. If the estate don't want you in there, they can simply lock the door, which some do but most don't. It sounds like we won't be welcome at the Black bothy; this will be a guerrilla visit. I have visions of Martin and me dressed in khaki with blackened faces crawling through the heather. Our visit will be a tiny stand against the landed rich men who own most of Scotland

and treat it as their plaything. For one night at least we will reclaim the bothy!

Martin appears from the kitchen bearing two enormous plates of full English, the consumption of which is definitely unwise for two gentlemen of our age, making it all the more appealing. As we both settle down for our appointment with fate, Martin looks up, knife and fork poised. 'Success?'

'Oh yes. Capital.'

At least we know the Black bothy is not a figment of Google's imagination as we walk in across the moorland. I've taken an obscure route in to avoid passing the estate office and alerting them to our presence. The moorland we are walking on is typical of those used for driven grouse shooting. It is a vast, treeless expanse of rolling country where the heather has been kept artificially short by burning so that the red grouse can prosper. Here and there are lines of shooting butts, little shelters about chest high, made of turf or mounds of peat. The butts are where the guests hide with their shotguns, waiting for the hapless grouse to be driven towards them by beaters. The grouse are little more than living targets and are often shot in their thousands in the name of sport. The shooters – or guns, as they are known – often have an assistant who loads the shotguns (they usually have a pair) so that they can keep up a high rate of fire. It's a landscape devoid of life apart from the grouse and any mountain hares who have managed to avoid being shot. On the map, there is a track heading over the shoulder of the hill and down and across the glen to the bothy. The track turns out to be a cartographer's daydream, and soon we are both plodding through the heather.

Martin is a few metres ahead of me when he stops and points across the glen. 'That might be it there.'

I step forward and can see a building half a mile away. Just as I am taking out the map to check its position, there is an explosion of sound and feathers at my feet. A grouse flies into my face, its wings beating the air furiously as it achieves vertical take-off. At the same time, it fills the air with its strange cry that is somewhere between a cough and a laugh. I stagger backwards in surprise, dropping the map, my heart rate somewhere in the thousands. Given the sudden shock and the fried breakfast earlier, I'm surprised my heart didn't pack in and leave me stone dead in the heather. I curse the bird as it flies off. I would never shoot a grouse, but sometimes I'd cheerfully strangle one when they ambush me with their incredibly annoying habit of waiting until you almost step on them before rising in total panic and making a determined effort to fly up your trouser leg. Considering the number of times they have done this to me, they must have shortened my life by several weeks.

Fifteen minutes later, we are standing before the Black bothy.

Martin recalls yesterday's debacle. 'Do you think it's locked?'

I examine the door. There's no padlock or keyhole. Seconds later, we are inside, dumping our rucksacks gratefully on to the wooden floor and hauling out stoves and coal. It's a small bothy and, as you might have guessed, it's black. It's largely made of wood with only one gable end. There is one room, which has a long rough table, a few chairs and a fireplace. Many of the bothies I stay in are old shepherds' houses. They are made of stone and have some age and character. The bothies on grouse moors are different. These buildings are not too old, perhaps seventy years, and no one has ever lived in them. They are constructed by estates as places where shooting parties can stop for lunch when they are tired of blasting grouse out of the sky.

Martin sets off exploring. 'My God, look at the size of these!' He has found a shelf with a collection of enormous champagne bottles on it.

He turns one of the bottles round so he can read the label. 'Moët & Chandon. I've never seen bottles of anything this big.'

'I think they call them jeroboams. Must have had them for lunch. Any of them full?' I ask hopefully.

Martin carefully inspects the tall green bottles standing to attention like soldiers on parade. 'No, all empty. It's a miracle they could hit anything after these.'

I begin to set the fire, always my job on our bothy trips.

Martin heads out to a stream to fetch water with his collapsible container, but he is back moments later. 'The water's no good. Come and see.'

I peer into the small stream that runs beside the bothy. Its water is as black as old sump oil, so full of peat as to be undrinkable. 'Oh bugger. It's miles back to the river.'

Martin points to a long wooden pole set across the stream, about two feet above the water. In the centre of the pole is a wire cage about the size of a shoebox. 'What's that?'

'It's a trap. I think they call them rail traps. They are there to catch stoats and weasels. That's what the shooting community calls "vermin".'

'I thought they were wildlife?'

'One man's vermin is another man's wildlife. If you call an animal vermin, it sounds fine to kill it.'

Martin shakes his head in disbelief. 'Why would they do that?'

'This is a grouse moor. Any animal that might predate on grouse is seen as a bad thing.'

Martin plods off with his water carrier and I go back to setting the fire. He returns half an hour later with a container full of peat-stained but drinkable water. By this time, as the outlines of the hills sink into darkness, I have the merry yellow flames dancing cheerfully in the hearth. We light the candles and the world shrinks to these four walls and Martin and me. Outside the bothy, the wind eases and the night sky clears. Soon the bothy is glazed with hoar frost, and fine ice crystals glitter on the fringes of the black water stream.

Later, as we sit before the fire, I talk about the moor around us. 'It's unnatural for this moorland to have to fight nature to keep it there. Truth is, to sustain a grouse moor with enough birds to shoot, then you have to keep the numbers of grouse high, and that attracts raptors like eagles and hen harriers.'

'Why are hen harriers are so badly hit?' Martin asks.

I explain that it's particularly difficult for the hen harrier because it is one of our furthest-ranging birds of prey. The harriers may live many miles away, but because they travel such long distances from their roosts to find prey, they are likely to discover grouse moors on their range and be drawn to them as hunting grounds. As a result, the pressure for gamekeepers to eliminate the harrier threat is enormous – indeed, the estate's economic survival might depend upon their killing harriers. This is, of course, illegal, but the land is so remote and the likelihood of being caught so low that the temptation to shoot or poison birds must be almost overpowering. Few keepers are apprehended, and the penalties for such activity have until recently been light. The great injustice when such men are prosecuted is that the real responsibility lies with the landowners, who perpetuate a system that forces their employees to break the law so that grouse moors can be sustained.

Martin looks up and casts his eye around the bothy, the empty champagne bottles catching the firelight. 'I'm not sure I like this place much.'

'This bothy?'

Martin drains his mug. 'No, this moor.'

I can understand how he feels; there is something about grouse moors that makes me feel uneasy too.

'Yes, you're right. This place reeks of death. Let's not come here again.'

Now at least the woman I spoke to in the estate office can rest easy in her bed. I won't be returning. The following day, Martin climbs on to the southbound train at Inverness station and rides off towards more confrontations with confused rail officials as he battles the absurd conformity of our age. Where would we be without the Martins of this world, who simply refuse to accept that carrying a lethal weapon on a train is no longer acceptable?

***

The end of January is almost upon me and I realise that my quest to see some of Scotland's iconic creatures has been far from successful. I have watched stags rutting and been eyeball to eyeball with a mountain hare, but so far I've not come close to seeing a pine marten or a wildcat. My interest in beavers was sparked after my conversation with Ben at Scottish Natural

Heritage. I still can't quite believe that there are beavers living in the broad farmlands of Tayside. All my life I have associated them with the great vastness of the North American wilderness. My childhood imagination was filled with shoot-outs with villains in forests full of wolves, bears and beavers. The image I have of beavers is indelibly etched with them as exotic creatures of the wilderness. As a result, I have difficulty accepting that they might be perfectly at home living in Tayside, just off the A9. Oddly, I don't have the same problem with lynx. I think that is because I had no childhood images of them. Lynx didn't show up in the gunfights of my childhood, so I don't have the same preconceptions.

I won't be satisfied until I've seen a Scottish beaver, so I head down the A9 once again, this time on the two-hour drive to Dunkeld. The Highland village looks about as far from the kind of place you might find beavers as it is possible to get. It's tiny, with only one proper street and a small square. It occurs to me that the square would be an ideal place for a shoot-out, but I try not to think about that. My childhood imagination has never really left me. Sadly for Dunkeld, the village has succumbed to a tsunami of bus tours and tourists. Its shops are full of Scottie dogs dressed in tartan and jars of whisky marmalade, something no Scot has ever spread on his toast. The place has become totally given over to removing money from tourists. None of the shops in its car-choked main street sells anything that a local resident might actually need. Dunkeld has become as much of a real place as Blackpool seafront.

I hurry on through and drive up the steep hill out of the village and then two miles further to the Scottish Wildlife Trust reserve at the Loch of the Lowes. There are two hides here overlooking a beautiful loch edged with woodland and teeming with wild birds. Swans, surely the king of all waterbirds, silently cruise the surface of the loch. The visitor centre has a glass wall, on the far side of which virtually every species of tit squabbles over nuts and seeds. I spot a yellowhammer and the woman who staffs the desk points out a nuthatch. Part of me wants to spend time here improving my knowledge of birds, but I am on a mission to find evidence of the existence of beavers. I confess to the woman behind the counter of the visitor centre that I am here to meet one of the exotic animals that populated my childhood imagination. She is sympathetic and points this old cowboy to a loch in the hills two miles away called Mill Dam.

It's a mild day as I walk up the track through the forest, and the air carries the smell of distant rain. The SWT reserve was lovely, but I am not a visitor-centre kind of person. I resent being guided to walk in a certain direction and view things in a particular order. I always get an urge to leap over the fence and make a break for freedom. As I climb, the soothing scent of trees, moss and leaf mould fills the air, and the staccato drumming of a woodpecker echoes through the trees. I emerge from the woods and a delightful wee glen opens up below me. The view here is very different from the rugged Highland landscape I am used to. Here I am at the very edge of the high country of Scotland. This is where the ragged hills of the north tumble into the broad, fertile plains of lowland Scotland and the populous central belt.

The merging of two landscapes is reflected in the wonderful mixture of language in the place names on my map. The Gaelic of the north is still evident in the names of some of the hills.

Here is Creag nam Mial and there Creagan Ruathair. The hill I am walking on is called the Knock of Findowie. There are places with names like Butterstone and the Mill of Muckly – surely they are populated by small people with large hairy feet who speak with tall pipe-smoking wizards. I like to think so. The landscape is peppered with low hills, with few rising above 1,500 feet. Between the hills, there is a network of small paths and tracks with occasional houses, where further north there would only be open moorland and space. I spend so much time in the hills that I sometimes forget that there is another Scotland far from the lonely places I frequent. Lowland Scotland has its own charm, different from the rugged hills I am used to.

Mill Dam is a small loch, perhaps 300 metres long and 100 metres across, surrounded by trees, with tracks around both sides. This Saturday, it's busy with dog walkers and cyclists. Not the remote quiet place I had imagined. There is, as you might have guessed, a man-made dam at the end where I am standing, so I begin to explore there. In minutes, I begin to see beaver signs. There is a small silver birch with its trunk half eaten, and all around are the classic cone-shaped stubs of young trees felled and carried away. As though to convince myself of the reality of these creatures, I reach out and touch the gnawed tree trunk. I have travelled through time. I am touching something that has been lost from this island for over 400 years. This is the first mammal to be reintroduced. My fingertips trace the bite marks in the trunk as a man in pink Lycra jogs past. It all feels so matter-of-fact, as if beavers have simply slipped back into our landscape and left hardly a ripple. I know it wasn't like that. There was a great deal of negotiation, and the beavers' presence has caused conflict with farmers. Despite that, these teeth marks on the side of a birch tree signal something momentous: they tell us that little by little the wild can return. We can find space in this crowded island for the creatures we have lost. I am less than three miles from the crowded high street of Dunkeld, with its tourists and shortbread, yet I am surrounded by the marks of an animal that was once lost, its presence on this island all but forgotten.

I spend the afternoon wandering around the loch, noting signs of beavers everywhere. There are saplings severed and removed, substantial trees toppled and left stripped of their branches. These signs of wanton destruction are the beginning of a process – the beavers will transform this landscape over the years and develop a wetland that will benefit other creatures and us. These are not subtle creatures. They leave unmistakable evidence of their presence, as if they arrived and erected a huge sign with the words 'WE'RE BACK!' emblazoned across it. These are not secretive lynx who would melt into the forest never to be seen. Beavers are the marching band of wildlife. They come into the environment with horns blaring and drums beating. Despite their obvious presence, I've yet to see one. I know that they are mostly nocturnal, so I must wait till dusk.

As the light begins to fail, the dog walkers, joggers and cyclists cease to appear and I am alone beside the waterside. I've found a track that leaves the water's edge and heads to where saplings have been gnawed. I slip behind some trees and wait. The temperature soon begins to drop and I pull my jacket tight. I now know that waiting in the cold and the dark is pretty much the staple activity of wildlife-spotting. I watch until I can no longer make out the edge of the loch, but nothing moves. Soon it is completely dark, and I fumble in my bag to find my head torch for the walk back to my car. I have resigned myself to the fact that I won't see beavers tonight. Then, from somewhere in the darkness, I hear a splash. It's different from the sounds of the loch in the daylight – the dogs lunging in for sticks or the ducks fighting in the centre of the lake. Although I cannot see it, I know instinctively it is a beaver. No one has heard the splash of a beaver's tail in these hills for 400 years. He is a time traveller returning to his lost homeland.

FEBRUARY

February opens like an enraged toddler in the throes of a tantrum. Britain is battered by high winds and torrential rain as named storms hurl themselves across the country one after another. Homes are flooded in Cumbria and yet again places like Hebden Bridge are inundated with water. We are reaping the fruits of neglecting our earth. Bereft of trees and criss-crossed with drainage ditches, the land can no longer soak up the outpourings of storms. Instead, our rivers swell to bursting point and flood our towns and villages.

Later in the month, Martin and Joe will arrive for their traditional midwinter long weekend. We have been meeting up at this time of year for over twenty years now, and have plans to head north for a bothy trip. I have a week or so before their arrival and I want to make the most of my time, even if the weather is conspiring against me. The nightly TV headlines of flooding bring home to me the importance of the beaver, and this makes me even more determined to try and see one of these animals for myself. I know my chances are slim this time of year, with beavers spending most of their active hours in the cover of darkness, but I have acquired a secret weapon, a camera trap. I don't have the skill or the patience to be a wild-life photographer, but even if I am not lucky enough to see a beaver I would be delighted if I could film one. The small camouflaged camera is more than just a technological toy – it is one more way for me to engage with the landscape. Through it, I hope to be able to glimpse the wild creatures that tiptoe past the bothy door while I sleep. It is another way of looking. The more you are prepared to look, to take time with the land, the more you will see and the deeper your relationship will become. I wish everyone was happy to hunt with a camera lens.

The Scottish beaver reserve of Knapdale sits a few miles outside Lochgilphead in the wild land of Argyll. The weather forecast shows a brief lull in the incessant storms, so I head off to drive the four hours from my home to the reserve. Driving down the side of Loch Lochy, watching the wind whip the waves into spinning vortexes of spray, it occurs to me how big Scotland is. Many folk say how much the world has shrunk, how jet travel has brought once-distant places only hours away. Yet I can still drive for hours through a landscape that is largely empty. The population density of the Highlands is around eight people per square kilometre. Statistics, of course, tell great lies, and even that is a gross distortion. Most of the population of around 240,000 folk live in the U-shaped area surrounding the inlet of the Moray Firth, which encompasses the ever growing Highland capital of Inverness. On my drive to Knapdale, I pass through a few villages and only two notable towns, Oban and Fort William. Sutherland, in the far north, has a population density of just two people per square kilometre and, last time I checked, they weren't speaking to each other. If you travel to Sutherland on a wet and windy Thursday in November, you could be forgiven for thinking that the only people who live there are a handful of grumpy shopkeepers who would rather you didn't disturb them by wanting to buy something. In the words of Billy Connolly, 'There's nobody here.'

I drive through the town of Oban, its quayside busy with black and white ferries shuttling to the Scottish islands. Now I am leaving the Highland hills behind and driving on into Argyll. The land becomes lush and green, a place of switchback roads and gently rolling hills. I turn off the main road a few miles before the village of Lochgilphead and on to a narrow but uncharacteristically straight road that cuts across a strange flat grassland and heads for the wooded hills of Knapdale across the Crinan Canal.

Arriving at Knapdale, I find myself in a narrow glen with a string of small lochs leading to the sea. I step out of the car, stiff from the long drive, on to the gravel of a small car park. I have arrived with perfect timing at 3 p.m., and it will be dark in about two hours. I am hoping to see beavers around dusk, which is when they start getting up and thinking about breakfast. The sky is full of grey, threatening clouds when I step out of the car. The wind is gentle by the standards of the early winter of 2020, which means I can let go of the car without worrying about being blown into the nearby loch. Yet there is still a stiff breeze and the tops of the tall trees surrounding the loch sway in the wind. I gather my camera, bird book and binoculars, and set off to look round the reserve. Today I've ditched my red hillwalking gear and I am dressed in my David Attenborough green jacket, hoping it will allow me to blend in with the scenery.

I see very few signs of beavers. I'm annoyed I didn't do more research before coming here, or at least read the notices in the visitor centre. Walking round the loch, I find only a handful of trees showing the teeth marks of these mammalian engineers. It's a short walk to the bird hide, where I spend a few minutes shivering and watching the water before the cold forces me to move on. Then I follow the short path that leads away from the loch, wondering if I might find beavers beside the heavily wooded burn that runs into the loch. There's a sign that says something about 'seeing the view'. I don't normally pay any attention to signs, but as it isn't far to the viewpoint I decide I'll be a good tourist for once and walk there along the wooded path.

I find myself on the apex of a small ridge, and the sight that greets me leaves me stunned. Below, the narrow ribbon of the Crinan Canal weaves its way to the west, where Loch Crinan opens up to the sea. The glittering water mirrors the grey clouds that rush across the sky and is broken by a patchwork of vivid green islands and peninsulas dotted with tiny white houses. Inland, a vast area of wetland opens up below me. The tall grass ripples with the wind sweeping in from the coast to lap against the feet of the low forested hills. The high ground rises abruptly from the flatlands and separates it from Loch Fyne away to the south-east.

The land at the foot of the hill I have climbed is broad and flat. The narrow road I drove along to get here is straight as a gunshot with no obstacles to make it bend. The plain below is the ancient Moine Mhor, the great bog. This is an old landscape, a huge deposit of peat that has been forming for over 5,000 years. This is one of Europe's last large raised bogs, and trapped within it are layers of history. It is a good day to see such a place. The wind races across the open land of the Moine to reach the sea, where it scars the water white with its fingernails. I sense a presence here. I am looking not only at the landscape but into the bones of time. The fact that the view has surprised me makes me see it more clearly. I have to make sense of the unexpected. Like suddenly meeting a stranger out on the long road, I want to see what stories this place has to tell. I see it not only with the light and shade of this winter's day, but with the bright eye of my imagination. Today, the wind is holding the landscape in its fist and shaking it – there is a sense of memory here.

This landscape has changed little over thousands of years. Once I would have looked down from here and watched herds of elk grazing in the bog. Around the edges of the bog, where the water drains into bright streams, beavers would have created wide pools and deeper wetland. The elk and deer would have raised their heads every few mouthfuls and turned their eyes to the ragged treeline. They would have searched the woods for a flicker of grey movement that might betray the wolf pack slipping closer. Sometimes a chase would begin, with wolves fanning out, pursuing elk across the grassland. If you were to dig down, you would find that memory held in the peat.

I stand for twenty minutes until I notice the light beginning to fade. It is time to head off in search of beavers. I will have time to visit the bog tomorrow. I reach the second loch just as the wind is beginning to rise and the woods are filling with impenetrable darkness. To my surprise, a woman holding the hand of a blond boy around three years old emerges from the forest. I wonder if they might have any beaver information but I am a little reluctant to approach them. I don't want to frighten them, as it is almost dark and this is a remote place.

In the end, I try and put on my least threatening manner. 'Hello, you haven't seen any signs of beavers, have you?'

The woman smiles, clearly not intimidated, and points to a track into the woods. 'If you head down that track, there are beaver signs there.'

She goes on to explain exactly where to find a tree that has been cut into by the animals. I follow her instructions and sure enough I can see where they have been stripping the bark from the base of a large tree. It is the perfect spot, as beavers always travel down watercourses and there is a deep ditch just a few feet from the tree leading to the loch. I take cover and wait. The darkness comes in quickly on this cloudy night and I realise that my chances of seeing beavers are slim. I can only hope to see them at dusk, when they begin to forage but there is still enough light. However, night falls so quickly that it is dark in less than twenty minutes. I hadn't allowed for that. I'm sure that's a mistake David Attenborough wouldn't make. My apprenticeship as a naturalist is taking longer than I expected. I wait alone in the woods, hiding near the tree until it begins to rain, and I wonder if this is the sort of thing a respected author should be doing.

It's time to deploy my secret weapon, the camera trap. There is a handy fence post close to the tree, so I strap the camera to it. As I am fiddling with the straps in the light of my head torch with the rain – by now icy and torrential – running down my neck, I make a decision: if I ever have the chance to live my life over again and do something different with it, whatever I choose to do, I won't be a wildlife cameraman. I'm only now beginning to understand just how much patience that job must require. Walking back to my car through the dark wood, I feel a little sorry for my camera trap all alone out in the rain, but rather it than me.

I spend the night in a modern bunkhouse on the shores of Loch Fyne. There is only one other resident, a Glaswegian man with missing teeth and a lot of tattoos who is running about the lounge stuffing his possessions into black polythene bags.

He looks at me with an expression of alarm. 'Ah'm oot of here. You no hear the radio?'

I confess I haven't.

He sucks his remaining teeth. 'Ah drive a milk tanker. Gonna heid back to Glasgie. Bad storm coming. There'll be trees.'

I stare at him blankly. 'Trees?'

He looks at me with an expression that tells me he thinks I am an idiot – not only do I not listen to the radio, but I don't know what trees are. 'Aye, trees. They'll be doon all over the road.'

While he's packing his bags, he tells me he heard that the storm will arrive about noon tomorrow, but he's taking no chances and leaving now. Then he talks of all the times he's been trapped in places I've never heard of by trees blown down in gales. By the time he leaves, I'm distinctly apprehensive about tomorrow's weather. I remember passing a great many very large trees on my way here, and the thought of them crashing down as I drive by is not appealing.

Next morning, I leave the bunkhouse early, hoping to make it home before the worst of the storm arrives. It is only just daylight when I return to the reserve to collect my camera trap. The rains have ceased, but the clouds are fleeing across the sky as if running from the coming storm. I follow the path back through the woods and down to where I set up the camera. It is sitting just as I left it, watching the tree with the beaver teeth marks on it. When I flick open the small screen, it tells me that the camera has taken several videos. I feel a surge of excitement – I'm about to see my first beaver. I watch each video expectantly and at last I see an animal moving in the bushes beside the tree. There is a movement in the branches and I am disappointed as a roe deer steps nervously into view. He wanders around the tree and then strolls over and curiously sniffs the camera, calmly checking out this new thing in his territory. It's not the beaver I had been hoping for, but it is a glimpse of a wild creature. Even though roe deer are common here, I feel some sense of satisfaction at having caught something. I have a lot to learn about deploying my camera trap, but at least it is a new challenge and something I am able to enjoy in the wild even when the weather forces me to stay low. I'm beginning to wonder if I jinxed the weather by calling my book *Wild Winter*. This winter has been a lot wilder than I expected.

I don't want to leave this area without taking time to have a closer look at the Moine Mhor, and so stop in the small wooded car park halfway across. Now I can see that the surface of the bog consists of tussocks of grass and moss separated by pools and black areas of exposed peat. The area forms a huge dome I had not noticed before. Now, in midwinter, it is quiet, and only a solitary buzzard sweeps through the darkening skies searching for prey. It must be very different here in summer as it's clearly an amazing habitat for wildlife. I make a note to return to Argyll – I must spend more time exploring this beautiful forgotten corner of Scotland.

Storm Ciara dumps a month's worth of rain on Yorkshire and Cumbria, deluging fields and flooding streets. The Met Office issues hundreds of flood warnings. Even the mainstream media begins talking about the potential benefits to flood prevention from reintroducing the beaver. There would be sacrifices to be made and we would lose some farmland and have to grow food somewhere. The beavers would have to be managed to some degree and it would not be a relationship without conflict. Although our island is such a crowded place, surely if we were to make some compromises we could live alongside these animals to the benefit of all of us. Perhaps we are finally beginning to learn that the planet will heal itself if we work with it and not against it.

***

A week later, I am standing by the window in my flat in Inverness. From my perch on the fourth floor, I can look across the River Ness to the narrow road bridge known locally as the Black Bridge. A few folk are walking across it into town, mostly shop workers heading in to start their day. I am watching for the arrival of two familiar figures, Martin and Joe.

Other acquaintances from my schooldays have faded away, but it's different for the three of us. Over the years, we have forged our friendship in the mountains, and bonds formed in these rugged places are strong. Together we have endured endless miles across rain-soaked hills. Martin and I ploughed through the knee-deep peat of the Pennine Way. Years later, Joe and I diced with death on the ice-wreathed cliffs of Ben Nevis. Together we have experienced the lows of cold, exhaustion and countless drenchings. We have felt the elation of exploring unknown remote landscapes and reaching the summits of long-imagined hills. The mountains of our lives have forged the connection between us.

After a few minutes, two unmistakeable figures appear on the bridge, Martin beneath his giant rucksack and Joe carrying a more modest pack, lacking Martin's compulsion to carry everything he might ever need in any eventuality. This is the sign I have been waiting for. I rush into the kitchen and prepare to unleash death by breakfast: bacon, eggs, sausages, mushrooms and beans. Joe and I also have black pudding. If the towering inferno of fat doesn't kill at least one of us, then my role as host has not been fulfilled.

The smell of bacon has already filled my flat by the time I hear the steady thud of Martin's size 11 boots approaching up the stairs.

'Ah, that smells good,' Martin declares as he enters, swiping a photo of Ben Nevis off the wall with Excalibur as he tries to turn in my narrow hallway.

Joe hangs up his jacket. 'All right, John?' When you have known each other over four decades, there's no need for initial pleasantries.

Martin is, of course, wearing his 1974 hill gear of woollen trousers and pullover topped with balaclava. Joe is more conventionally dressed in jeans and a fleece, and at least looks as if he might have a nodding acquaintance with the twenty-first century. He now lives in the Borders not far from Ayr and joined Martin's train at Glasgow. Joe shares Martin's suspicion of new technology and neither of them owns a mobile phone. That's what has happened to folk my age: they can be neatly divided into those who at least attempt to master new technology and those who regard it as some evil form of witchcraft.

I wait until Joe sits down at the table. 'Tea?' I know he'll want tea, but I suppose I have to ask.

Joe nods, places the science fiction novel he was reading on the train on the table, and fondles the knife and fork expectantly. Martin joins him and, as I put mugs of tea in front of them, reaches into his pack for his small plastic sugar container. That's the benefit of knowing me so well. He knows I won't have sugar, so doesn't ask and brings his own. He always carries a stock of sugar in the white plastic pill bottle his father gave him forty years ago from his chemist's shop in Bebington.

'So you got here, then?' I call from the kitchen as I crack the eggs into the frying pan.

The weather map has been a patchwork of yellow warning areas over the past week, with rail networks telling customers to avoid travelling if at all possible. At one point, we had considered postponing our annual get-together, but had decided there was no guarantee that better weather was on its way and the pair might as well take their chances and head north.

'Only forty minutes late,' Martin announces triumphantly as I place the plate of fried offerings before him. 'I knew if I got past Carlisle I'd be okay.' One useful by-product of Martin's mathematical mind is that he has an encyclopaedic knowledge of train timetables and the inner workings of rail networks. This means that if one train doesn't run, he'll know every possible combination for getting to his destination on time.

Martin drenches his breakfast in sauce. For him it has to be brown sauce, specifically HP. Preferably in a glass bottle, but plastic will do – he's not quite as obsessive-compulsive about his sauce as he is about other things. Beans must be Heinz, and when I suggest he might like to try the low-sugar variety, the look of horror on his face tells me everything I needed to know.

Peanuts have to be KP. I won't go on. Martin has a food list which I think was probably programmed into him when he was about ten. He will only eat items on the list. Bacon, sausages and eggs are on the list, so he eats those. Items such as black pudding and salad are not on the list and therefore cannot be eaten.

As well as being an avid science fiction reader, Joe collects obscure facts. He picks up the HP sauce bottle and examines the label carefully. 'Did you know that HP sauce comes from Holland?'

I look at him stunned. 'HP sauce comes from Holland?'

Joe swallows some bacon. 'Says so on the bottle. Look.'

'But that's Houses of Parliament sauce. It's British.' I feel as if I've been lied to. I don't even like the stuff, but HP brown sauce has been around all my life, and you knew that the very bottle you held in your hand was also having its bottom smacked by an MP over some bacon in the canteen in the Houses of Parliament. Now it turns out to have been foreign.

I shake my head in disgust. 'Remember Harry Shackleton's shop in Bebington village?'

Joe nods mid-sausage. 'It's a hairdresser's now.'

'When I was a kid, we used to get a monthly grocery order from him. Without being asked, Harry would put a bottle of HP sauce in our order. He knew my dad's sauce consumption, so he'd pop the bottle in as ours got low. Tesco don't do that, do they?'

Martin peers at the bottle in disbelief. 'You'd never think it was Dutch. It's got the Houses of Parliament on the label!'

Joe pursues the last of the sugar-rich baked beans around the plate. 'Won't be able to get them after Brexit. There'll be a sauce importation quota.'

Martin shakes his head in despair. 'Oh Gawd.' He can't eat bacon and egg without HP sauce – he might starve.

Joe finishes his tea. 'How's the book going, then?'

This is a painful subject. 'Well, I was ill most of January. I haven't managed to see beavers, whales or wildcats and the weather has been lousy. Apart from that, it's fine.'

Martin laughs. 'Well, it has been a wild winter then, hasn't it?'

Joe is practical as ever. 'It was a bit ambitious hoping to see all those. What *have* you seen?'

I think back to October and November. 'I saw the rut, I got really close to a mountain hare and it's still only February. I've got until May, when winter actually ends in the high hills.'

Martin puts a sauce-doused sausage to the sword and looks up at me expectantly. 'Well, what's the plan?'

The plan has been exercising my mind for some time. Every few days, some meteorological phenomena called Charlie or Gladys or Humphrey lands on us and fills the streets of Cumbria and the houses of Hebden Bridge with a year's worth of rain, or blows enough wind at us to flatten every tree in Scotland. Even though our modest aim is to spend a night or two in a bothy, the tricky question of which bothy has been troubling me. Only a few years ago, we were trapped at the wrong end of Loch Arkaig when torrential rain flooded the loch and swamped the road and marooned us. Fortunately, it stopped raining and the loch went down as fast as it had risen, but that was a lesson learnt. Inver hut, with its walkway across the bog and proximity to the road, would usually have been my first choice, as we could have done some hillwalking from there. However, I remembered my recent paddle through the bog there and decided against that as a reasonable choice. Being the old man of the mountains has a downside when your decisions can make the difference between a successful trip and a dismal failure.

I run through all the options I can think of, then something occurs to me. 'Most of the bad weather seems to be missing the far north.'

Martin shovels home the last mushroom. 'That's true.'

'I know a bothy that has a short, sheltered approach with no rivers to cross. I reckon we could get into it in pretty well any conditions.'

Joe sits back in his chair, groaning from breakfast. 'Sounds like the one, then.'

***

The following day, Joe and I are driving north. Martin has elected to catch a train half the way there and then get on the little minibus service that will take him to our meeting point in the village of Durness. His love of public transport baffles me, but it makes him happy and it's his holiday after all. Joe and I drive through the small village of Lairg at the eastern end of Loch Shin. It has a couple of shops, a petrol station and even a bank, all huddled against the eastern end of the loch. That may not sound a great deal, but it is the last community of any size before we head north into the emptiness of Sutherland.

'This is where civilisation ends,' I tell Joe as we pass the shop and post office.

We drive on for twenty minutes and only meet one car. The weather has been so wet that the loch has burst its banks and flooded the surrounding woodland, with the tops of trees standing forlornly in the water.

From the passenger seat, Joe gazes out to the distant hills. 'It's a wild place this, isn't it? That's some expanse of water.'

'Aye, it's something like seventeen miles long.' I'm watching the sky darkening and the grey clouds building above us. 'Looks like the weather's getting rough.'

The sky is raked by great black clouds piling one on top of the other. Two or perhaps three miles away, across the far side of the grey water, a huge plume of water shows brilliant white against the dark sweeping hills. It is cascading down from the hillside and thundering into the loch with immense power. We are exposed to the wind here and the car rocks menacingly. One of the dark clouds collides with us and sends bullets of sleet and hail hammering into the windscreen. Double-speed windscreen wipers have no effect and I am reduced to crawling along the road blinded by sheets of rain. At last, we leave the shores of Loch Shin and head north to meet a much smaller but no less impressive body of water, Loch Stack. The wild weather brings a grandeur and savage beauty to this area. A mountain, Arkle, rises sheer above the waters of the loch. Here the surrounding hills and the wooded lochside give us some shelter from the weather.

'I've been thinking,' Joe announces. 'Do you know I've still never been to the Cairngorms? Maybe after this there'd be time to pay a visit there?'

'I suppose we could.' There would be enough time in the four days Martin and Joe plan to stay if we headed directly there after this trip north. 'I know a good bothy we could go to.'

'Great. I've always wanted to go there.'

Then I remember Martin's views on the Cairngorms. 'You'll have to talk Martin into it. He's not a fan of the place.'

'Why not?' Joe's surprised.

I shake my head sadly. '"Great humpy things", he calls the Cairngorms.'

Joe knows that once Martin has made his mind up he is not easily moved. 'I'll have to work on him, then. Maybe he'll come round.'

Loch Stack is normally calm and I often stop here to take photographs of this spectacular place. Today, the storm rakes the loch and whips up waves topped with white foam. I won't be stepping out into the maelstrom. Soon we reach the point where Loch Stack empties into a small river which runs for some four or five miles through the steep-sided valley to the sea.

'See that river? That's the Laxford, one of the best salmon rivers in Scotland.'

Joe's not an angler, but this sparks his interest. 'Have you fished there?'

'You're joking. I asked one of the locals I met in the pub at Durness if I could chuck a fly in. He said, "Jesus, you cannae fish there! The Duke of Westminster owns it. They'll bloody hang ye!"'

I pull over into a small lay-by fifty metres from the river. 'Come and see this.'

The rain has eased a little as we pull on our cagoules and head towards the swollen river. There, only feet away, is a lovely fishing hut with a small veranda overlooking the water. As an inveterate bothy-hunter the thought of spending a night in this idyllic spot fills me with excitement. The door is locked.

Joe peers through the window. 'There's a little log-burning stove and even a sink.'

'I'd love to spend a winter's night in there.' I can see myself sitting warming my toes, dram in hand and all the world at my feet.

Joe ponders the idea. 'I'm sure a good bottle of malt whisky, placed in the right hands, would get you the key to that door.'

I laugh. 'Yes, I'm sure you're right, but I've no idea which hands to put it into. I got Jon Barton, the boss of Vertebrate Publishing, to write to the manager of the estate and ask.'

'Nothing to lose, I suppose.'

'Jon said that one of his eminent authors would like to spend a night here. Part of my research on a new project.'

Joe convulses with laughter. 'I hope he didn't say it was the bloke who wrote *Sky Dance* as an attack on the landed gentry. That would definitely get you hung. Did he get a reply?'

'No, of course not. They wouldn't want some oik like me in there.'

We climb back into my car and head north, the wild scenery of Sutherland unfolding around us. We drive on in silence for a while, then Joe turns to me. 'I've been thinking … '

Joe is always thinking. Outwardly he appears quiet, but inside his brain he is always considering possible scenarios and potential new technologies. This is one of the reasons he's such an avid reader of science fiction. He sees endless possibilities.

Joe continues, 'Do you know you can turn seawater into hydrogen and run cars off it?'

I wonder if he's gone mad. 'Can you do that?'

'Oh yes,' he assures me. 'Can't understand why we don't do it in Scotland. We've plenty of hydroelectric power.'

I'm still not convinced this isn't some delusion. We are men 'of a certain age' and minds do snap from time to time. 'Wouldn't there be a danger it would blow up?'

'No more than petrol.'

We drive on to Durness in silence while I consider Joe's sanity. Durness lives its life in a constant battle with the wind. It's as if God is angry with the place and is doing his best to blow it off the face of the earth. An icy wind jostles the car as I come to a halt beside the white-painted shop and post office at the heart of this village. Martin, incredibly, has made it here by train and minibus. He is hiding from the wind in the rickety bus shelter, his itchy 1970s woollen balaclava pulled down over his ears. I had thought that this form of headgear, universally worn by hill folk in those days, had become extinct. I now know that Martin has a small colony of them surviving in his wardrobe.

Durness is typical of many remote places that are at the end of the line for travellers. On the east, Wick and Thurso are such places. On the west, Ullapool and Kyleakin, where the bridge links Skye to the mainland, are similar. They attract unusual people who, for one reason or another, find mainstream life uncomfortable. Such people find a train or a bus that will take them off into the sunset and get on it. When their chosen form of transport stops, they get off and take root, often for no other reason than because they can't go any further. Thus Durness has two types of residents: people who have always lived there and can see no reason to leave, and those who found themselves washed up on the beach there with all the other driftwood. Sometimes the people who drifted there are seeking something like a better life away from the rat race. At other times, they are escaping from something, trying to bury themselves somewhere so remote that no one will ever find them. The mistake in this is, of course, that no matter where you go, your demons always travel with you.

When I get out of the car to meet Martin, the wind tries to wrestle the door from my hands. I knew it would be like this and so have a tight grip on it. There is a great little campsite nearby called Sango Sands, on a field that leads down to a wonderful beach. In summer, the place is almost idyllic, so it might surprise you to learn that in winter you can stay there for free. The reason for this is that in the winter you would need to anchor your tent with telegraph poles to stop it blowing away. If you got up for a pee in the night, the only way to prevent yourself ending up in the Atlantic would be to crawl to the toilet block holding on to the grass for grim death.

Martin's narrow range of menu options means he has to plan ahead for each meal in case he's faced with something that isn't on his list. 'What's for breakfast tomorrow?'

'Porridge!' I yell over the wind.

Martin shakes his head and heads for the shop, emerging a few minutes later with a small tin of what he describes as 'b and s', or beans and sausage. Porridge, you see, isn't on the list.

While Joe nips to the small public toilets, I quickly google turning seawater into hydrogen. It turns out this is a thing. Joe's sanity is no longer in doubt.

When he returns, we begin to divide our supply of coal into smaller bags ready for the walk in to the bothy. Joe looks up from the bags of the black stuff. 'You remembered candles, didn't you?'

I have to confess I didn't and we both go into the shop. Candles are becoming hard to find. Look in any supermarket and you'll find hordes of short stubby candles that smell of everything from sandalwood to an Arabian brothel. These candles are designed to smell and emit a dull glow. They are not there to produce light. Hardly anyone uses candles for light. Only we bothygoers who return to the Dark Ages on a regular basis use candles for light.

Joe looks on a shelf. 'Here we are.' He holds up a bag of tea lights.

'I'm not using those.' I am a sworn enemy of the humble tea light. 'I can't understand why people insist on taking tea lights into bothies despite the fact that they give off virtually no light. And, for some odd reason, people also think they burn. I've dug thousands of empty foil tea lights out of the ashes of bothy fires.'

Joe is surprised by my passionate outburst against these innocuous stubby candles, but dutifully replaces them on the shelf.

Then I remember. 'There's another shop we could try.'

Joe looks at me suspiciously. 'There's two shops here?'

I nod. 'Oh yes.'

Joe shrugs. 'I'm amazed there's even one.'

I have always wondered how such a tiny place as Durness managed to support two shops; now I am about to find out. We drive a quarter of a mile to the second shop. I knew it was there but have never been in it. The post office and grocery store we have just visited is a veritable hypermarket compared to the second shop, which is tiny. Martin and Joe have decided that I'm wasting my time trying to buy candles here and remain in the car, safely out of the wind.

I push open the door and an ancient bell jingles. The shop is smaller than my front room. Behind the counter is a smart middle-aged woman who is talking to another woman of about the same age. The shopkeeper smiles at me, but the pair continue their earnest conversation. Stacked on shelves around me are a few odd cans of food and bottles of beer, enough to justify defining the place as a shop but not enough to bring folk here in any numbers. While the two women talk, I stand wondering how on earth this can be a commercial success. Then it dawns on me. This is a perfect idea. The shop doesn't sell groceries; it deals in a much more valuable commodity: conversation. The lady behind the counter is simply using her front room as a shop. Her friends come in, buy the odd thing just for appearances, and then they have a nice chat. It's a brilliant business model. She has virtually no overheads and when hordes of visitors turn up in the summer to drive the North Coast 500, she might even make some money. After a few minutes, the conversation comes to a close. The 'customer' leaves and the lady behind the counter turns to me as if she has all the time in the world – because, of course, she has.

'What can I do for you?' she says.

'I was wondering if you had any candles?'

She thinks for a moment and then, to my amazement, takes a pack of proper non-scented candles from the shelf behind her. 'We have these.'

'Oh these are great.' I reach out to take them but she draws back. Perhaps I'm too hasty, maybe I should have spent ten more minutes talking about the weather.

'I'll just give them a wee dust.' She carefully wipes a thick layer of dust from the box of candles. They have probably been there since the miners' strike.

I offer her a ten pound note.

She smiles indulgently. 'I've no change.'

Of course she has no change – she never sells anything. I retrieve a pound coin from the car and the transaction is complete. For some odd reason, I leave the shop with the feeling that I've enjoyed myself.

Half an hour later, we are tramping towards the bothy. At first we are exposed to a wild wind, but after a few minutes we enter the shelter of a small wood. This was part of my plan. I hoped that, no matter how foul the weather, once we got into the wood we would be protected. I know there are no sizeable rivers to cross, so unless we get struck by lightning we'll make it to the bothy. I have visited this place several times over the past few years and I am always amazed by the incredibly luxuriant lichen in this wood. The trunks of the trees and their twisted boughs are swathed in a thick growth that almost looks like leaves. I don't recall seeing anywhere else where the trees are so well covered. Even in the height of summer, the air in this wood is damp. Its proximity to the edge of Loch Hope and the constant rain-bearing westerlies that sweep in from the coast must make it so wet here.

The bothy is a small white cottage at the waterside, only a few feet from where the wind sweeps the water into white-horse waves. The gable end, which faces the worst of the weather, is cracked, and the edge of the roof is tattered from the wind. A few slates have come loose, but considering it must be two years since I came here last and this winter's storms have been ferocious, this simple shelter is remarkably intact.

There is something in these timeless places that calms the soul, far away from the cacophony of everyday life. I am drawn to the stillness of remote locations. On this peaceful lochside, the waves beat endlessly against the shore, as they have done for thousands of years and as they will do for thousands of years after my footprints have been washed from the sand. Last time I came here, I followed the stream that runs beside the bothy back up the hill until I came to the crumbling walls of an old cemetery. Many of the old headstones leant at odd angles and some had fallen flat. The wind and rain had long since washed the names from the stones that still stood. They were a reminder that lives were once lived here.

Soon the shadows are becoming long as the day grows weary and the sun dips behind a distant ridge. An hour or so later, all three of us are fed and sit bathed in candlelight as the coal in the hearth spouts yellow flames. Martin peers at the fire, willing it to rise up. 'My feet are cold.'

'Really? I'm fine,' I tell him and give the reluctant fire a poke.

'It's not that warm,' Joe concurs from the far side of the fire, his misting breath confirming the temperature.

Years of winter bothy visits seem to have rendered me more impervious to the cold. Perhaps it's simply that over the years I've refined my choice of clothing until I'm perfectly adapted to bothy life. Martin is swathed in two of his beloved old pullovers. Joe has a fibre pile jacket and jeans. I have a large down jacket, the sort of thing one might wear in the Himalaya. A recent addition to my kit has been a pair of tartan fluffy trousers, which are every bit as elegant as they sound. They cost me five pounds in Lidl, and are as light and warm as they are ridiculous.

I'll settle for warm and ridiculous over smart and cold any day. Actually, I secretly enjoy looking ridiculous – it's one of the few things I am good at. The room slowly warms up as the wine and the whisky go down.

I reach in to my rucksack and produce the camera trap. 'Seen one of these before? I got it recently to try to film wildlife.'

Joe takes it from me and instantly opens it up to examine its inner workings. 'It's movement-sensitive, is it?'

'I think it's more to do with sensing different temperatures. It picks up things that are warmer than the ambient temperature.'

Joe's analytical mind is working. 'So it's no good for reptiles, then?'

Martin normally avoids anything technical but is puzzled by this. 'Why not reptiles?'

That thought hadn't occurred to me. 'He's right. Same temperature as everything else.'

Joe laughs and closes the camera's case. 'So snakes are invisible. Let's go and put it up.'

Martin thinks about coming out with us. 'I'll stay in here.' I don't think he likes the idea of invisible snakes.

The moon is drifting over the loch, sending its bright reflection out across the waves. We cross the small stream and catch the scent of the fire smoke from the bothy.

'What about over here?' Joe has found the perfect spot.

A small higgledy-piggledy stone wall runs down beside the stream almost to the loch's edge. It leaves about a three-foot gap through which all men and animals must pass. I strap my camera to a post and then deploy my secret weapon: a small sachet of cat food.

Joe laughs. 'I saw that earlier. I wondered why you'd brought cat food.'

'I'm hoping that the scent might bring an otter out of the water.'

I press a switch and a little red light blinks as the secret watcher springs into life.

'When are we going to ask Martin about the Cairngorms?' Joe asks as we follow our head torches back to the warmth of the bothy.

'He won't want to go.' I'm not optimistic of persuading Martin. 'He hasn't changed his mind since 1974.'

Martin is toasting his feet by the fire when we come back into the candle glow of the bothy.

We join him at the fireside and Joe begins. 'We've been thinking … '

Martin looks up from his wine suspiciously.

'How about going to the Cairngorms after this?' I suggest gently, finishing Joe's sentence.

Martin's face clouds. 'Why do you want to go there? We could visit another Sutherland bothy.'

'I've always wanted to go to the Cairngorms,' Joe announces.

Martin pauses and wrinkles his nose. 'Why? I can never see the appeal.'

'We could go to Glen Feshie. Lots of rewilding going on – it would be worth seeing.'

Martin looks unconvinced, and I know I'll have to think of something to sell the place. 'There's a flash bothy. Even has a toilet.' Martin has never been comfortable with using wild facilities and likes bothies with toilets.

'Hmmm.' Martin pulls his woollen pullover tighter as he reluctantly considers the proposition.

Then Joe slides something into the conversation. 'Of course, you could take the train to Aviemore. And you've never actually got off the train at Aviemore, have you?'

This is a master stroke.

For Martin, another opportunity for rail travel and a never-before-explored station is an irresistible combination. He visibly brightens. 'No, I've not actually got off there before. That could be interesting.'

Joe will get his first visit to the Cairngorms.

During the night, a violent storm comes in. The bothy shakes and hail bounces off the roof so loudly it rouses me from my sleep. It's good to feel so warm and comfortable when there is a maelstrom only a few feet away.

In the morning, after our porridge, Joe and I retrieve my camera. The cat food has vanished, so something has been there. Joe and I eagerly scroll through the captured footage. The first is a red deer sniffing curiously at the camera just like the roe deer did at Knapdale.

'I never knew deer were so curious. I suppose it's not surprising that they would want to check out anything new,' I tell Joe.

The next image captured – or not captured – is the animal responsible for eating the cat food. Unfortunately, it came in under the angle of the camera.

Joe shares my disappointment. 'You must have set the cat food too close to the camera. Listen – you can hear it wolfing down cat food, but can't see the animal.'

This is something else I am learning. 'If it was an otter, we'd have seen its back. It must have been something smaller – a weasel, or possibly a stoat.'

The stone wall would be perfect habitat for either of them. Close but no cigar again. Mastering my camera trap is taking me longer than I thought, but it is fascinating to see animal behaviour I didn't know existed.

There's one more image left to look at on the memory card. Perhaps my cat-food eater is on that. The screen shows nothing at first, then the infrared shows a bush moving gently in the breeze. Then we watch as a pair of distinctly human legs walks through the camera's field of vision.

'Who the hell is that?' Joe asks, echoing my thoughts.

I read the camera time on the video. '21.04. Pitch dark by then. We never left the bothy. There's nowhere anyone could have been going at that time of night.'

Joe thinks for a moment. 'Perhaps it was someone coming to the bothy who didn't want company and decided to go home.'

This seems plausible until we realise that if the night walker had turned round, the camera would have caught him leaving. We watch the video again in silence. Neither of us can explain this mysterious figure.

'Let's not tell Martin,' I whisper to Joe.

Joe nods. 'We might never get him to a bothy again.'

***

The Cairngorm mountains are full of legends. Perhaps it is because they are so vast. You can wander for days and see no one, find yourself lost in oceans of impenetrable mist and witness storms of unimaginable ferocity. Even for our ancestors, used to living amongst wild things, they inspired wonder and fear. There are the well-known tales, like the Grey Man of Ben Macdui, a tall, lonely figure walking alone across the slopes of that huge hill in the emptiness at the heart of these mountains. The Grey Man is the Scottish yeti, the Bigfoot of Badenoch. There are other stories too, like Finn MacCool tracking the last wolf across the hills to its death or the Red Hand of Glen Mhor, a ghost tale that has almost been lost. These all owe as much to the florid imagination of storytellers over the years as they do to reality. Truth is a commodity that does not grow stronger with time.

There is one modern legend that I hope is true and is worth the telling. It is about Glen Feshie and how the fate of this place was changed by one man. The glen was purchased in 2006 by the billionaire Anders Povlsen. It was his intention to develop it as a traditional shooting estate with a conference centre and golf course included. With the investment he was able to make, this Highland glen would become as prestigious as any in Scotland. Povlsen brought his staff together to unveil his plans and there was a lot of enthusiasm about the proposal as they shared drinks from silver trays afterwards. Not everyone was enthusiastic, however. The estate manager, a man called McDonnell, had a different vision. He approached the billionaire and asked quietly if he could spare ten minutes to go out in McDonnell's Land Rover to see the estate. As they drove through the dark forest and out on the open hill, McDonnell spoke of his dreams for the glen. He spoke of bringing the wild back, of breathing new life into the empty valley. He spoke of the trees returning, of the glen bursting into life once more, of restoring the birdsong and the glen becoming the lush place it once was. McDonnell must be a persuasive speaker and Povlsen a great listener, for in that short drive the new owner changed his plans. He decided to rewild the glen and give Scotland something far greater than a state-of-the-art conference centre: he would give Glen Feshie back its wild heart. All of that may be true or it may be false, but I hope it becomes a modern Cairngorm legend. It deserves to be a story that is told because it speaks of hope, and we need that more than ever now.

***

Martin, Joe and I huddle around the map. It glows in the darkness, illuminated by our head torches.

'We can't be far away,' I announce, trying to sound confident.

Joe grunts, in a sort of way that tells me he wants to be warm and dry. 'You said that half an hour ago.'

He's right. I cast my head torch out into the darkness, hoping to find the angular corner of a wall or the reflective glint of a window, but all I can see are trees and huge wet snowflakes falling silently through the darkness.

'We must be at this path junction.' Martin points at the map to where two tracks meet. 'And the bothy should be there.'

I struggle to see the lines on the map. Even with reading glasses, I'm finding the detail of maps harder to read, but I am too stubborn to admit it. Martin is right. If we are at that path junction, then the bothy should be just over there. There's only one problem: it isn't. We are all tired. The coal in my rucksack is getting heavier by the minute, yet we seem no closer to enjoying its heat. I have a growing suspicion that we are lost due to my incompetence. I try not to let Martin and Joe know, although I think they have worked it out for themselves.

Martin examines the map again. 'Are you sure it's not that way?'

I'd been in the direction he indicates and discovered a distinct absence of bothy. 'It's not there – I've looked.'

'Well, let's look again,' Martin insists.

Joe shrugs and we all march off in the direction I've already visited. We walk for another fifteen minutes and, just as I predicted, there is no bothy. We are just staring at a wall of darkness.

I turn to Martin to chastise him for not believing me. 'You see, there's nothing … '

'What's that?' Joe is standing about a yard further up the track. I hurry to where he stands and my head torch beam picks out the corner of a stone building. Seconds later, we are crashing through the door.

This is Ruigh Aiteachain, a legend in bothy-going circles. It is a two-storey stone bothy that has been extended to provide room for a staircase to take you to the top floor. I assume that there was a ladder before that. No expense has been spared in its construction. The windows are double-glazed and the floor appears to be oak. The bothy is topped by a brand new metal roof, and the whole place is as warm and watertight as a bothy could be. There are two rooms on the ground floor, both with stoves. This is deluxe.

An hour later, we are fed and relaxing around the stove when the window is illuminated by head torches and a couple in their mid-twenties rattle their way through the bothy door. The man is tall with angular features and a neatly trimmed beard and introduces himself as Mark. His partner, Anna, is also tall with her long blond hair tied back in a ponytail. Both of them look lean and fit.

Martin clears a space for them on the table. I cooked our meal and I have an uncanny knack of being able to occupy all available space.

Mark unpacks his rucksack in a well-organised manner. 'Tricky place to find in the dark. Did you guys have any trouble?'

'Oh no, we found it no problem.' My ego as a mountain author speaks for me.

Joe coughs.

Martin raises an eyebrow.

I'd better change the subject quickly. 'So where are you guys from?'

The couple are from Cornwall. Mark is working towards his Mountain Leader certificate, while Anna works in a hospital. This is their first time in the Highlands and they are loving it.

'There's so much space up here,' Anna beams.

Mark glances around the candlelit room. 'This bothy's brilliant. Are they all like this?'

Martin chuckles. 'Er, no, this is the best one I've seen.'

'Have you been to many?' Anna asks as she examines Martin's 1970s woollen hill outfit.

Martin nods at me and grins. 'He's been to all of them.'

'Not quite all. But this place is the Ritz of bothies. It's downhill from here,' I tell Anna.

Joe is ever practical and points to the floor. 'This floor alone must have cost thousands. I've got a kitchen extension at home in the Borders and I couldn't afford a floor like that.'

There is a story I heard around a bothy fire. I'm not sure how accurate it is. 'Well, I heard that Povlsen came in here when it was the old, run-down bothy. And he's supposed to have said that he wouldn't expect anyone to sleep somewhere he wouldn't, so he spent £250,000 doing the place up.'

Joe stands and inspects the windows. 'I can believe it cost that. The windows are double-glazed, custom-made.'

There's no doubt this is a wonderful bothy and great care has been spent on its renovation. I'm pleased to see that the old door has been retained upstairs. It's an unusual door, low with a curved top. I always thought that it made the place look a bit like a hobbit house. It's great that the estate have put so much effort into it, but I feel slightly uneasy here. The bothies I am used to are basic buildings. They are often well maintained, but retain their character. Somehow this place has been sanitised. It's clean, warm and dry, but it's lost something for me. I am more at home in the rough, draughty bothies of Sutherland than I am behind the double-glazed windows of this super bothy.

Mark unpacks the couple's gas stove, empties a packet into a shiny pan and stirs in boiling water.

Martin eyes the contents suspiciously. 'What is that he's eating?' he whispers to me.

I peer surreptitiously into the pan and the mixture inside looks vaguely familiar. 'It's the same stuff that bloke was eating on the Pennine Way. The one who was plugged in to Facebook.'

Martin thinks for a moment, going through his memory index. 'Oh yes. What did he call it now? Couscous!'

'Yes that's it. It's what all these young backpackers eat.'

Martin is unimpressed. There's no way this strange stuff will replace the beans and sausage he's been eating for the past fifty years. 'I don't think we ever found out, did we?'

'What?'

Martin shrinks away from Mark, who is now spooning the stuff on to his and Anna's plates. 'What is couscous?'

Mark pauses. 'I'm not really sure what it is. Anna's vegan. I'd rather have sausages, but she makes me eat it.' Mark gives us a sly wink to show us he's not serious.

Anna sighs. 'I don't make you eat anything. If you really want to eat meat, you can do what you like.'

Mark gives a laugh and begins tucking into his meal again. 'No, I'll force it down.'

This is obviously a constant source of good-natured banter between the couple but we still don't find out the origin of couscous. There are some mysteries in life for which there are no answers.

Anna looks up from her steaming meal and pushes back her blond ponytail. 'What do you think about this virus, then?'

Joe has been sitting quietly with his science fiction novel, but looks up at this. 'Could be serious. It's coming our way.'

Mark sighs. 'I think we haven't got long before it kicks off here.'

I try to lighten the mood. 'Matt Hancock says the risk to the public is low.'

Everyone bursts into laughter. I can't think why.

An hour later, everyone has had a share of whisky and we are all relaxed in each other's company.

Martin is curious about the couple's journey. 'Where else in Scotland have you been?'

Mark's eyes brighten with enthusiasm. 'Glen Coe, spectacular place.'

'And then Glen Shiel. We wanted to come here to Glen Feshie most of all, though.'

'What do you think of the Highlands?' I ask them.

The couple are silent for a moment, as if reluctant to speak.

'It's beautiful,' says Mark.

'But it's so barren, isn't it? Overgrazed, I mean,' Anna adds quietly, as if worried she'll offend us.

In her twenties, she can see what it has taken me forty years to understand. 'Yes. Yes, it is.'

Suddenly there is a passion in Anna's eyes. 'It could be so different, so alive with creatures. And yet it's dead.'

We spend the next forty minutes discussing what has happened to our mountain environment. I'm not only surprised by how strongly she holds these views, but by her knowledge and understanding of the environmental challenges facing us. I am ashamed to say that she is far more knowledgeable than I am. At her age, I knew nothing of these things. I accepted the mountains as they were and didn't question how they got there or what was happening to their wildlife. I suppose I was typical of my generation. Listening to this couple, I realise that they are part of the next generation of outdoor folk. For them, the hills are more than simply views to admire or challenges to overcome; they are part of a living, breathing earth that they have a sense of responsibility towards.

'We wanted to come here to see a valley where things were different, where nature was being given a chance,' Anna ends by saying.

Joe laughs. 'And I wanted to come here to finally get to the Cairngorms after forty years.'

I suddenly feel thirsty after all this talk. 'Who fancies tea?'

I fill my old aluminium pan with water and set it on my stove. I light the gas and the room is filled with a comforting roar.

Mark peers at my ancient aluminium pans curiously. 'I've not seen pans like those. How long have you had them?'

Suddenly, I am conscious of how battered my pans look. The many dents and scratches they carry bear witness to their years of service. Searching through the mists of time, an image comes to me of a young man with long hair and tatty RAF trousers – me as a university student – standing in a shop in the Lake District. I take a set of shiny pans off the shelf and buy them with the money I was given by Harold Wilson's government to get beer with. There was no suggestion that I pay the money back, ever. It was a present. Thanks, Harold. The Kinks had just released 'Schoolboys in Disgrace' and Jimmy Savile was on TV with *Jim'll Fix It*. No one had heard of the internet and I'd never seen a computer.

Mark smiles, waiting for me to answer. I look down at my pans. They are veterans of a thousand bothy nights and have rattled their way across the Alps, spent nights under the stars on Mount Kenya and shivered with me in the Rocky Mountains.

How can I tell this bright-eyed young man that my pans have spent twice as long on this earth as he has? 'Er, quite a while. Maybe twenty years.'

Martin is thinking, searching his memory banks for the origin of my cooking pots.

He nods. 'They've lasted well.'

Like the young man who bought them all those years ago, they look like they have lived a bit. Their shine has gone, like my hair. They are not quite the same shape as they used to be, and neither am I. As I think about all the meals these pans have cooked, I recall the scares that there used to be about aluminium causing dementia. At least that wasn't true. Then it occurs to me that if it were true and I am demented, I'll be the last person to know.

Martin looks up in triumph. 'I remember now, you had those pans on the Pennine Way. That's forty-four years ago.'

He goes on to tell Mark and Anna about the hell of the Kinder Scout peat bogs, and we laugh as we recall finding a Dutch walker sunk to his waist and having to haul him out by brute force. We pass a happy hour recalling the misfortunes of our youth.

Later that night, Joe and I make our way out into the forest a hundred metres or so from the bothy. It's late now and the clouds that brought the snow flurries have passed away, leaving the sky starlit and the heather jewelled with frost.

We strap the camera to a tree overlooking a likely animal track and head back to the bothy.

'Let's hope there's no passing zombies tonight,' Joe laughs.

'Yes, one was enough.' I try to sound casual but can't help scanning the forest with my head torch before we return to the warmth of the bothy and crawl into our sleeping bags for the night.

***

In the morning, Joe and I rush out to see if my camera trap has captured anything noteworthy. It's a cold morning but the ground is soft; the frost of last night has already thawed. In the daylight, we can take in the bothy properly. It sits snugly beneath the brand new roof, its stonework immaculately pointed. Cobbled paths have been constructed around the walls and one leads to the toilets fifteen metres away. The small building is surrounded by trees, buried in the forest. This is one reason why we had difficulty finding it, as my map shows Ruigh Aiteachain sitting in a clearing. This is testimony as to how much the forest has regenerated since the deer numbers were reduced. Looking at the bothy, I am reminded of the gingerbread house Hansel and Gretel discovered deep in the forest in the old fairy tale.

Joe is staring into the trees. 'Where was it we put the camera?'

I remember strapping the camera to a tree last night, but I am struggling to remember which tree and there are a lot of them. 'I'm not too sure.'

It's now that I learn perhaps my most important lesson in setting up a camera trap: it's a good idea to remember where you left it. It had seemed so obvious last night, but now that I see small trees everywhere it's impossible to recall which one we chose. The camera is camouflaged, which won't make finding it any easier.

After we have looked for twenty minutes, Joe calls to me. 'Here it is!'

When we search through the images on the camera, we are disappointed to find nothing new. Wildlife, I am beginning to realise, is consistently uncooperative. The camera did detect something as it shot two short videos, but these turn out to have captured only grass waving in the wind. The instructions warn that this can happen if an animal like a deer runs through the camera's field of vision. The camera takes around half a second to fire up, by which time the deer could easily be long gone.

Back in the bothy, Mark and Anna are packing to walk a circuit over the hills and back to their car at the road at the start of the glen. They have already been for a quick jog through the forest and are talking excitedly about the regeneration they have seen and the richness of the glen and what a wonderful place Glen Feshie is.

'So, Anna, do you think you'll come back to Scotland?' I ask.

Anna grins back at me. 'Of course – there's so much to see.'

Mark is packing his rucksack. 'We want to come back in the summer. Maybe Sutherland – we hear it's wild up there.'

Martin laughs. 'It certainly is. We've had some wild weather there.'

Anna and Mark head off on their walk over the hills and our older trio sets off on a more leisurely exploration deeper into the glen. Only a short distance from the bothy, we find ourselves walking through old twisted Scots pine, descendants of the Great Forest of Caledon. The ancient forest once stretched all the way up from the Pennine spine of England on through the Cumbrian hills and the wilds of Northumberland to finally expand into the vast wastes of the Highlands. The forest was home to lynx, bears and elk. It hid the Celtic people who had made it their home. The Scots pine trees that remain here fill the air with their scent. They smell of wild places, of a time long ago when these hills were truly wilderness. These trees bear the marks of their wild lives. Their limbs are twisted and contorted from a thousand storms, and the landscape of their bark is buckled, riven by the harshness of many winters. These trees are survivors, warriors from a different age.

Martin stops to put a roll of film in his camera, and Joe and I wander amongst the broad trunks of these magnificent trees, which exude a sense of timeless peace.

Joe explores the reddish-brown bark and deep grooves of a nearby tree. 'These Scots pine look well battered. How old do you think they are?'

'I think I read somewhere they can live for up to 300 years.' It's odd to think that the tree Joe is touching might have been a sapling in the early years of the Industrial Revolution; staggering to think what changes have occurred in its lifetime.

Martin takes some photographs to add to the already enormous archive he has collected during his lifetime and we walk on. Soon the River Feshie splits into two courses. Here the glen widens and is full of light. From the trees comes the sound of birdsong. Blackcaps, bullfinches and tits flutter amongst the trees and bring the air to life.

Joe stops and gazes across the landscape. 'Well, I made it to the Cairngorms at last.'

I join him in looking out across the open floodplain of the river. It is dotted with small clusters of trees, which colonise places high enough to avoid the rage of the river when it rises in spate. This is a great flat expanse with the hills rising up beyond.

Joe continues. 'It's the sheer scale of the place, isn't it? It's so big.'

'Yes, it is impressive.' Martin has joined us, and even he is being won over by the magic.

Further on up the glen, it narrows once again and the new growth presses in around us. The trees are reclaiming this area which once was theirs alone. This is a living glen, full of trees, birds, mosses – bursting with life. I cannot stop looking and listening and absorbing. My imagination is full of possibilities. What will this place be like in ten years, twenty years or even thirty years? There could be beavers and lynx here, and perhaps one moonlit night a wolf will howl across this glen.

Joe takes in the scene. 'Look at all the young trees. You never see growth like this in the hills. All from reducing deer numbers.'

Martin clicks away with his camera.

I feel something exciting and vibrant here. 'This is how Scotland could be; this is how it should be. Let's hope this is how it will be.'

These trees were once condemned to be the last of their race, the new growth eaten away by the innumerable deer sustained for shooting estates. Things are different now. Beneath the branches of the old pines, young trees are growing. They are only the size of Christmas trees, some no thicker than my finger, but they are the vanguard of the returning forest. They are the giants of the next generation. They show so clearly that if you give nature a chance, she will take it. All across Britain, we have stood with our boot on the throat of nature, only allowing her the slightest chance to breathe, but now things are changing.

'You know what we need now?' I laugh. Joe and Martin turn to me. 'Beavers in the Feshie.'

I can only imagine how quickly beavers would change this landscape and create areas of wetland all along the course of the river. Things are changing in the Scottish landscape. Now there are beavers in the Tay and otters in the urban rivers of Edinburgh. In Glen Feshie, there are young Scots pine bringing back the memory of the great forest. I have walked for so many years through empty, silent glens, through hills burnt barren by the gun. I have spent so much time hand in hand with the stillness of death that to see a place like this, so full of life, fills me with joy. I am glad that McDonnell took his employer on a drive through the glen and told him how special it could be, and I am even happier that Povlsen listened.

This has been a good weekend. Joe has finally seen the Cairngorms, Martin has managed another trip to make the most of his retirement and maybe I have learnt something. I have seen that there is a change coming in the way people treat the landscape. The old generation is giving way to the new – not only the people who own the land but the folk who wander it too. I had thought that my generation was perhaps the last to want to travel these wild places. I know now I was wrong – we are not the last hillwalkers.

I think back to last night in the bothy. I recall the passion in Anna's eyes, and the knowledge and understanding she showed when she talked about this land. She is part of the next generation of hillwalkers and outdoor folk. Her generation knows and understands this landscape and the challenges it faces far better mine ever did. I have a feeling that as we walk towards the future, these hills and glens will be in safe hands.

MARCH

Martin's giant rucksack heads away down platform one of Inverness station and on to the 7.24 a.m. southbound train. Joe follows, happy that he has fulfilled his long-held desire to visit the Cairngorms. Our trip may even have convinced Martin that, far from them being big humpy things, there is something special about those hills. Back in my flat, the Radio 4 news is playing over my little transistor. I am only half listening as I drink my tea and gaze out across the rooftops towards Ben Wyvis in the distance. Something has happened over the last few days – the mountaintops are shining white with a deep covering of snow. Winter has finally arrived.

I listen to the news more intently these days. It is all about Covid-19. There are hundreds dying in Italy with people afraid to leave their houses. I listen in disbelief. Could it get that bad here? Surely we must be better prepared? There is another word I keep hearing: 'lockdown', whatever that means. I've been in denial that this virus will affect my life, but I am beginning to realise that there is something terrible coming. I want to explore the Highland winter and thought I had two months left, as the season can easily linger into May on the high tops. Now it seems I only have weeks or even days before all my plans are terminated by this sinister virus. I decide that if we are going into lockdown soon, I am going to make the best use of what time I have.

I begin feverishly planning. One of the things I had hoped to do was visit James Roddie's wildlife hide on the Black Isle, where I might be lucky enough to get close to a pine marten, a lifelong ambition. I phone him to find out my chances of seeing pine martens.

James doesn't sound positive. 'It's not great right now. She's visiting the hide about 3 a.m., so you'd have a long, cold wait and she doesn't always show. You might be better to wait.'

'Okay, James. Give me a call if she starts coming at a more sociable time.'

James had explained to me that pine martens, being creatures of habit, often call at his hide around the same time each night for two or three weeks, but then change their time for no obvious reason. The idea of waiting through a long, cold March night does not appeal.

So that's not an option. I put down the phone and stare vacantly out across the River Ness. The river is swirling and dark, full of meltwater from the hills. It rushes through the town, half a dozen miles from Loch Ness, to where it empties into the Moray Firth. Sometimes seals swim up the river to hunt fish in its lower reaches. More commonly there are ducks diving and feeding in the stony river bottom. I think the ducks are eiders and goosanders, although my fledgling bird-identification skills may have led me astray. They have perfected the technique of diving at the exact moment I manage to focus my binoculars on them. I then try to predict where they are about to surface, while they do their utmost to come up somewhere else. I end up catching brief glimpses of them and scuttling for my bird books to make sense of what I've seen.

There's a movement in the water. I can't tell if it's a seal or an otter. Even a small seal is much bigger than an otter, but out in the water it is difficult to judge scale. I run round the room in a desperate search for my binoculars and eventually find them sitting on the window ledge right in front of me. In the time it took me to find them, the animal has dived from sight. I focus on a patch of water where I expect it to surface. By a miracle, a head pops up right in the centre of my field of vision. I am still trying to decide what it is when the animal, its body as glassy as the water, dives once again. This time I catch sight of a flick of a tail as it sinks. It's an otter! I can't help smiling at the sight of the otter here in my flat with my slippers on and mug of tea in hand. I recall the frozen hours I spent beside Loch Aline and the Sound of Mull searching for these creatures. Then I saw no otters at all; now one swims past my house.

It's a large dog otter with a broad head. He's moving fast, swimming with the current towards the sea. Upstream of here are a series of small islands; I guess he's been hunting there. Now he is travelling in a straight line, perhaps back to his holt, which may be in some corner of the small nature reserve a mile or so from my flat. He prefers to travel underwater. The current is fast – the outpouring from thousands of hill streams combines here to form a powerful torrent. In half a minute he is carried away beneath the Black Bridge, a hundred metres downstream. I watch him dive for the last time with such an easy grace it seems as if his body becomes water. His tail flicks up, he sinks and is gone, slipping into the secret world below the surface.

I'm beginning to learn some important lessons about watching wildlife. The first is that wildlife doesn't cooperate.

It's rarely where it's supposed to be and often where it shouldn't be. The second thing I've learnt is that I was being wildly optimistic when I started this winter's quest. So far I still haven't managed to see a beaver, and the otter I saw was a lucky accident. My camera trap has only caught deer and ghosts. I have so much to learn. I certainly wouldn't call myself a naturalist, not even an amateur. Despite all that, this has been a fascinating winter. I have been wandering these Highland hills most of my life, yet I realise now that there is still so much to see and do. The urge in me to explore these hills and seek out the remaining traces of wildness is growing stronger. It feels like time is running out and, as well as seeing wild creatures, I want to visit the snow-covered winter hills. I want to catch my wild winter while I still can. I pull out a handful of maps, open them and spread them across my lounge floor. A world of possibilities unrolls before me.

***

The screen lights up on my phone and shows my location with a small red arrow. There I am, just passing the imposing iron gates of Balmoral Castle. They are three metres high and have Her Majesty's crest emblazoned on the front. These are serious gates, strong and forbidding, the sort of gates you don't mess with. They are closed and this old hillwalker, in battered hill gear and with a rucksack scruffy from years of use, is outside them, which is exactly where he is supposed to be. I've seen a lot of posh estates, pretty much all of them from outside the gates. It has to be said that this is number one. This is the poshest place in the whole of Scotland. It probably isn't the richest, but Her Majesty is at the top of the tree. When it comes to the tweed-clad, wildlife-murdering fraternity, she is the governor beyond all question. I tiptoe past the gates, hoping not to be savaged by corgis or arrested by the armed guards who are undoubtedly watching me on CCTV.

For forty-odd years, I have found my way in the hills with map and compass. My first attempts at map-reading were in the Lake District when I was a schoolboy. The maps that Martin spread out before me were a mysterious mass of swirling lines that bore little relationship to the landscape around me. Martin began my education, having some rudimentary skills his father had taught him on previous Cumbrian trips. Over the years, on the misty summits of Scottish hills and in the teeth of Sutherland's gales, my understanding of maps grew. I learnt the art of mountain navigation in the only way possible: by getting lost and finding myself again. I have taken pride in my navigational ability – after all, it is something hard-won. My struggle to find the bothy in Glen Feshie convinced me of something I had known for a while but had refused to admit: my deteriorating eyesight has undone my ability to navigate by map and compass. I have capitulated and now I am subject to the will of a GPS. I am learning the ways of the electronic beast. The little arrow on the screen tells me exactly where I am. I no longer need to find that mental connection between the lines on the map and the landscape around me.

'Here you are, old man. Right here. If I had legs, you'd be redundant,' my phone tells me, as another slice of life passes into history.

In my map-strewn flat, I had considered all sorts of possibilities, from travelling to the Isle of Skye to visit the Lookout bothy on its remote northern tip, to returning to Sutherland to visit the lonely Kervaig bothy near Cape Wrath. In the end, the lure of a night in a snow-bound bothy was too much to resist. If you want snow in a Scottish winter, the best place to go by far is the Cairngorms, so here I am lugging in coal to a remote bothy once again. Balmoral is in Deeside on the far eastern side of the Cairngorm mountains, and a good two-hour drive – over some of the most notoriously treacherous roads in the Highlands – from my home. I'm heading for Gelder Shiel bothy, which sits at the foot of Lochnagar at the heart of the Queen's estate. The little red arrow takes me up into the forest and then leads me out on to the open hillside and into a white world of drifting snow. Once I leave the shelter of the woods, I am exposed to the icy wind that sweeps down from the mountain and across the snowfields where there is nothing to stop it. The glen opens up wide before me as the arrow on my screen takes me along the snow-covered track. Ahead there is nothing but an open snowscape leading upward towards the iced pyramid of Lochnagar. The lone feature in the landscape is a stand of trees, black against the whiteness, a mile and a half away. It is this isolated patch of dark woodland I am heading for. I know that the bothy nestles in the shelter of the trees. In truth, I don't need to navigate, but nevertheless I check the GPS screen regularly to follow my progress and become familiar with my digital map and compass.

No one could call Gelder Shiel bothy charming. It is a practical grey box with only one tiny window in the rear. The bothy is a simple one-roomed affair, but it is wood-lined and cosy. There is a stove and a table and even a large supply of firewood supplied by Balmoral estate – thanks, Liz. On two bunks I find sleeping bags, and a gas stove and pans are laid out on the table. I light the fire and unpack my sleeping bag. Soon the bothy is warm and I settle in for the night. It doesn't take long for the warm fire and one or two whiskies to take effect. My eyes grow heavy and I doze comfortably in front of the stove.

At about 8.30 I wake with a start as one of the logs on the fire spits loudly. Looking around, I realise I am still alone; the owners of the sleeping bags have not returned. It has been dark for three hours now, and I am suddenly wide awake, worried that they may have encountered a problem high on the hill. There is a bitter wind when I go outside with my torch. I switch it off and search the hillside for any sign of returning lights, but there is only darkness on the vast mountainside. I do this every twenty minutes for the next hour. By 10 p.m. I am really worried. I open my phone, but there is no signal. I am just wondering how far I'll have to walk to be able to call for help when the bothy door rattles open.

A tall blond young man enters, climbing slings still hanging from his neck. He's followed in by a woman of about twenty. Her face is reddened by wind and she looks tired.

The young man grins. 'Oh, you've lit the fire. That's great. I thought I'd have to try and get it going.'

The girl sits down and begins to take off her snow gaiters. 'It's nice to get warm.'

'Are you okay?' I ask anxiously.

The young man laughs casually. 'Oh yeah. We did a climb. Took a bit too long. Had to finish in the dark.' They are students from Aberdeen and they have just completed their first winter climb.

The young woman is taking off her cagoule and beginning to look warmer. 'Only had one head torch. So we were slow.'

I'm thinking they must have been very slow when I remember how long I took on my first winter climbs. I offer them tea and whisky. They take the tea and decline the whisky. I keep asking them if they are all right and they keep telling me they are okay. I tell them I was worried about them and they burst out laughing. I am fussing over them like a father over his teenage children. They are fine; they don't need me.

I return to my seat by the fire. It turns out that their stove is empty and I am able to lend them mine. They have hardly any food, as they had been planning to finish the climb and walk out to catch the bus for Aberdeen. I've a spare tin of tuna and some soup that I give them.

'Here, take this in return,' the young man says, offering me a squashed cake.

I politely decline it. I am pleased to help them and give them what I have. I don't need the food and they do. Helping other people in the hills with no expectation of anything in return has been the code I have followed all my life. Perhaps it's me who looks as though I need help.

Sitting by the fire as the students cook their meal, I remember that when I was their age I'd have laughed at someone who worried about me. Perhaps becoming a father changed me forever so that I have an ingrained paternal instinct for younger people. Feeling slightly embarrassed at my concerns, I turn in for the night and leave them to enjoy the fire.

When I awake early the next morning, the students are still sleeping. I pack my rucksack as quietly as I can and leave them my gas canister and a small quantity of porridge. I can't let them starve, can I? The sky is just getting light as I close the bothy door. The chill wind of yesterday has freshened and carries with it glittering clouds of ice crystals. The bothy is festooned with a fringe of icicles almost a foot long. Last night's fire must have melted the snow on the roof and the meltwater has refrozen as it reached the cold night air.

There is a small cottage opposite the bothy. This is reputably used by the Queen to take tea on the rare occasion that one of the royal family isn't making headlines for some indiscretion. The snow has made some nice shapes on the roof and I am trying to get some arty photos of them when I spot a security camera watching me. I give the camera a cheeky wave, wondering if some secret service officer has been monitoring my every move.

Tonight I want to spend the night in another bothy on the far side of the Cairngorms to make the most of my time in case this lockdown actually arrives. As I pull on my rucksack, I notice that there are broken clouds racing across the sky and the wind is gradually rising. The weather is clearly breaking. My little red arrow and I head off across the snow. After a mile or so, I hear an engine, distant at first and then increasingly closer. A snowmobile cruises into view. Riding it is an SAS-type figure dressed in camouflage from head to foot with a rifle casually slung across his shoulder. I remember waving at the security camera. Perhaps that wasn't a very smart move in retrospect. I picture a group of angry soldiers somewhere in a command centre, watching me taunt them on the screen. 'Take him out, Ginger,' their officer barks.

Ginger roars to a halt beside me, switches off the motor of his snowmobile and grins from beneath his camouflaged cap. 'All right, mate?' He has a hint of an Australian accent.

'Er, yes, I'm fine,' I mumble, waiting for the difficult question about waving at the camera.

Ginger shifts the rifle on his back to a more comfortable position and looks up at the clouds. 'Weather's changing, mate. I wouldn't hang around if I was you.'

'No, no, I won't,' I stammer.

Ginger kicks the engine of the snowmobile into life. 'Have a good day.' With that he roars off across the snowfield.

He vanishes over the hill and the silence returns. I think that was the first time I've ever seen a snowmobile in use in the Scottish hills. Being convinced I was about to be shot by the SAS is a first too. By the time I reach the forest, I am glad of its shelter, and it takes less than an hour to get back to my car.

There is something odd about the road as I head back across the Grampians to Inverness. It is eerily quiet. I meet no cars coming in the opposite direction and my rear mirror shows only the line of tarmac snaking back across the hills. Then I turn left on to the road to the Lecht ski centre and there is a forlorn sign swinging in the wind: 'Road Closed'. That explains why I didn't see any other vehicles. I turn right and head into Morayshire. The landscape quickly changes as I drive away from the Cairngorms. Soon I am driving through low rolling hills of rich farmland that supplies the grain for the whisky-distilling industry. This is a huge area. If I had been able to drive direct to Inverness, I'd have covered 70-odd miles and arrived just after lunch. The way ahead now will hold no snow-blocked roads, but it is some 130 miles. That's the scale of the Cairngorms: going around them to get to the opposite side will mean a long drive.

Despite the long journey, I am still excited by this glimpse of winter. Even after all these years, a night in a snow-bound bothy is a special thing. The cold, the icy wind and the landscapes are magical things to experience. Ever since I was a small boy, the power of snow to create extraordinary landscapes has inspired me. This late in the year, the new snow will not linger long in the Highlands. Two or three days of rain could return the hills to a dull brown. If lockdown is coming, I want to make certain I am making the most of what little winter I have left. Perhaps I can squeeze one or two more bothy nights into this spell of frigid weather.

***

It's late in the day by the time I arrive in Aviemore and the street lights are coming on in the tourist-trap village. Aviemore was born of Britain's burgeoning ski industry back in the 1960s. Before that, all this place had was a station and a few houses. For a brief period, it was Scotland's St Moritz, the haunt of celebrities and what was known as the jet set. Now the village has lost much of its glamour and the ski development has fallen into disrepair, the victim of our warming climate in years when the snow fails to arrive.

I leave my car a hundred metres beyond Glenmore Lodge, the outdoor education centre, where the public road ends. From here, a broad track leads through the forest and up into the hills beyond. I shoulder my pack and head through the twisted outlines of the old pine trees as night begins to fall. I pass the small deep pool of the Green Lochan; its surface is stilled by ice and dusted with fresh snow. This evening, the hills are empty of other walkers and I am alone. The silence swallows me.

Half an hour later, I leave the forest and head out on to the open hillside, and after a few hundred metres my head torch picks out the rough walls of the bothy. It's dark by the time I arrive at Ryvoan and dip my head to pass through its low door. This bothy is more like the rough shelters I am used to than the deluxe bothy in Glen Feshie. It is an old stone cottage with walls that all look slightly less than perpendicular. Like Gelder Shiel, this bothy is a one-roomed affair. Unlike the Balmoral bothy, it lacks good insulation, and by the time I finish my evening meal I am beginning to feel the chill of the place seeping into my legs. As yellow flames begin to light up the dark interior of the stove, I wonder if my coal will last the whole night. The change in the weather Ginger spoke of has arrived now and I can hear creaking as the wind worries the corners of the roof.

This is the first bothy I visited when I began to explore these rough dwellings some ten years ago. I'm sitting drinking tea and wondering how many nights I've spent in places like this when I hear voices outside. Seconds later, two men of about my age enter the bothy. I can see by their rucksacks they are heavily laden.

'Hello, I'm Andy,' says the taller of the two with a shock of white hair as we shake hands.

His balding, stocky friend Rick groans as he sets down his rucksack. 'Tam will be here soon enough. He's a bit slow on his legs these days.'

There is an unmistakable Dundonian lilt to their voices as we talk about the weather outside and I make room for them in the cramped bothy.

Ten minutes later, Tam arrives, red-faced and sweating. 'Jesus, I'm getting too old for this.'

Andy laughs. 'What do you mean, "getting too old"? You're there already. I was about to head back down the track to see if you were deid.'

Tam shakes his head sadly and turns to me. 'You see the abuse I have to put up with?'

It's obvious from their banter that these three are old friends, and when they produce copious cans of beer and whisky bottles from their rucksacks I know at once what kind of evening this is going to be. The four of us spend the cold night heaping coal and wood on the stove. Despite our efforts, the icy wind outside ensures that the bothy never gets totally warm.

Soon the three are telling stories, mostly involving one of them falling in a river or embarrassing himself as a result of overindulgence in alcohol. Each accuses the others of being subservient to their wives. They tell me stories of their wild nights lost searching for bothies and I tell them some of mine as we watch the flickering fire and listen to the wind tugging at the walls. A generation ago, it was stories like these that bound the climbing community together. The tales were told around campfires or in the corners of Lake District pubs. Bothies are one of the few places that the old tradition of storytelling still survives. There are a few traditional storytellers in the Highlands, but their art is a dying one – television and the internet have all but killed it. Yet in the bothies of the Highlands, on odd nights when old men fall into company, stories come to life again. It is the sharing of common experiences that binds us.

I tell them that this bothy is situated on an RSPB reserve. 'There's a lek here, just the other side of the track. I was here in May a few years back and watched the black grouse dance.'

Andy looks at me blankly and quickly lightens the mood with another jibe at Tam for falling down the steps of the Queen's Hotel in Dundee. It's clear that black grouse hold little interest for the trio.

I am beginning to feel as if I have heard their stories before, and then I realise that some of them are the same ones I tell. Like Martin, Joe and me, they have spent a lifetime in the hills together and look forward to these bothy nights when they can recall adventures from thirty years ago. To them, that is more important than talk about what is happening in the wild land around us, about the birdlife and beavers. I fall silent and sit listening to their tales. I am still enjoying their company, but my perspective on the hills has changed.

This must have been what Anna and Mark saw when they looked at my friends and me in Feshie: three old men talking about their adventures in the hills long ago. We had done with them what these three are doing with me now, talking about memories of distant days. Perhaps one day Mark and Anna will talk about their first visit to the Cairngorms when they met three veteran hillgoers in a bothy. Will the two students I met after their climb on Lochnagar remember the old man who gave them tuna and left them porridge? These days we spend on the hills not only create memories but also define who we are. I suppose that's why they call it memory bank. In our youth we put memories in to our accounts, and as we grow older we take them out. That's what Mark and Anna will do when they are my age. Now that I understand this, I listen happily to the three Dundonians telling their tales. After all, they are only taking out the richness they put into their lives twenty years ago, and that is what nights in front of bothy fires are for.

***

The Dundonians are still snoring when I wake the following morning. I write my contact details on a scrap of paper, place it on the bothy table and leave quietly, just as the first rays of light are brightening the morning skies. I want to make the most of this day and visit another special place, Uath Lochans on the southern side of Glen Feshie. These hidden lochans are on the opposite side of the River Feshie from the bothy we visited last month, so we couldn't call there then. I pop into the supermarket in Aviemore and pick up a sandwich for lunch. To my surprise, a handful of people are wearing face masks. I've never seen anyone wearing one before. Getting back into my car, I decide there are bound to be some people who overreact.

As I step out of the car, the scent of the pine trees fills my nose and I feel at once a sense of peace. This is a special place – there is a magic here. I cannot see it, but I feel it in every pore. I leave my car and wander slowly through the huge arches of the trees; it feels as though I am walking through a vast natural cathedral. In a few minutes, I reach the edge of the first glass-still lochan and follow a boardwalk that leads out across the bog through a field of head-high reeds. The patchwork of small lochans surrounds me. Beyond them, the wind sighs through the Scots pine; above them, the snow-covered Cairngorm hills rise to merge into the clouds. These small areas of water were formed 10,000 years ago when the ice retreated, leaving great frozen blocks sinking into the earth like the tears of the Ice Age. There is a sense of the primeval here. Little has changed since then. The bear and the wolf have gone, and the elk with its huge antlered head no longer stands chest-deep in the water drinking, but the memory lives here.

The high reed beds are filled with birdsong. Beyond them is open water. The surface is quietened by a thin sheen of glass-clear ice that moves with the rippling water as if in remembrance of the great blocks. Out of the shade of the trees, the sun catches me in the open spaces. There is warmth in its rays now, a sign that the winter's progress is almost done and the riot of spring is marching closer. It is years since I last came here, and I wonder how I could have forgotten such a place.

Sometimes when I walk down a glen, turn a corner in a forest or crest a hill, I arrive somewhere I know is special. I can never predict it – that's part of its charm. It's that moment when I stand in silence struggling to comprehend where I am. It is the sense of the place that speaks to me, tells me of a connection deeper than my understanding. It came to me forty years ago when Martin, Joe and I walked in to the great hidden corrie on Ben Eighe. It happened when I walked to the lost bothy Poca Buidhe, and it happened when I stepped out of the door of Glendhu bothy near Kylesku and saw the mirror-calm sea loch. It happened ten years ago here when I stumbled across these lochans. It never leaves me, this sense of wonder at a place. It is a connection with the earth. I feel it now, listening to the wind in the trees and watching the still water. I am utterly content.

A few hundred metres through the forest, I come to a small wooden bird hide. I sit for a few moments and wait, watching the bird feeder full of peanuts rocking in the wind. After only a few minutes, a blue tit arrives with flashes of yellow beneath his busy wings. He looks at me suspiciously. I sit motionless. He grows bolder, takes a nut and flies off. Seconds later, he is back. He is more confident this time, and feeds on the nuts without feeling that he has to leave. A bullfinch joins him on the feeder. Emboldened by the little blue tit's presence, he feeds too, puffing out his reddish-brown chest. Then both birds fly off and are immediately replaced by two smaller crested tits with their punk hairstyles. The new arrivals dance about with the wind catching their Mohicans until the blue tit returns and bullies them away with his bluster.

Through my binoculars, I can see every detail of these tiny creatures. The blue tit weighs less than fifteen grams, but he struts about with total confidence, driving off the other birds whenever he decides they are in his way. My flat in Inverness has no garden, so watching these birds feeding is a delight I wouldn't otherwise be able to enjoy. I linger longer than I intend to, watching them squabble and fight over nuts like schoolboys at an impromptu party. There is so much to see. The wild weather of this winter has forced me to spend most of my time away from the high tops, which I had thought would lessen my enjoyment of my days in wild places, but that has been far from the case. I am finding so much to do and so many places to go as I struggle to learn about the creatures I see and the places they inhabit.

***

When I switch on my car radio, the announcer tells me full lockdown is only days away as a result of Covid-19. It is no longer a question of *if*, simply *when*. Social gatherings are banned and everyone is urged to follow social distancing, but the full force of lockdown has yet to happen. I realise that spending time close to folk in bothies, as I have just done, is not a wise thing, but I am tempted to get in one last bothy visit before this virus strikes. There is a bothy I know of deep in the hills and I've decided to try one last trip before that freedom becomes a memory. The single-track road winds its way further into the hills to the north of the Creag Meagaidh massif. I drive over bridges, through woods and beside lonely lochs, so far into the hills that it seems incongruous that there is a road here at all. This road simply finishes in the sort of place that anyone of sane mind would describe as 'the middle of nowhere'. It's as if the people building the road simply got to a point where they couldn't be bothered going any further. As I climb out of the car, I can see the bothy only a short walk away. Despite its accessibility, or perhaps because of it, this is one bothy I have never visited before. I have already decided that if there is anyone else in residence, I'll politely turn around, head back to my car and drive home to Inverness. Social distancing in a bothy simply isn't possible.

I load a bag of coal into my rucksack, and ten minutes later I am standing outside the small stone shelter with a battered green door. I am reaching for the handle when the door opens before me and a small man steps out. He is wearing a battered old tweed jacket, what they call a full Norfolk, the same as

Mallory wore on Everest. He sports a grey beard twisted into two spikes. These aspects of his appearance are striking enough, but what I really notice about him are a collection of badges he has running down the lapels of his jacket that jingle and glitter as he moves. Some of the badges are birds, the type the RSPB sell to raise funds. Others are various associations, including the Mountain Bothies Association. Another, strangely, is that of the Moorland Association, a group which supports driven grouse shooting. His dishevelled state and threadbare tweed suggest he is unlikely ever to have been on a grouse shoot.

He looks pleased to see me. 'I saw you coming down the road. Come on in,' he says in a southern English accent.

I don't go in, but I do lower my rucksack to the ground. 'No, I won't, thanks. The virus, you know.'

He looks puzzled – perhaps he is one of the three people on the planet who hasn't heard of Covid-19. Then the light of understanding shines in his eyes. 'Oh, you don't want to worry about that nonsense.'

He moves towards me as if to pull me into the bothy. I step back.

He looks disappointed. 'It's all a plot to control us, keep us from realising what's actually going on.'

I get the sense that asking him what is really going on may lead me down a very dark road, so I change the subject. 'Been staying here long?'

He thinks about this question and then begins counting on his fingers. 'Fifth night here, I think. Cup of tea?' He steps forward, more enthusiastically this time.

I step back even quicker, not because of the virus but because this bothy dweller emits the kind of odour that can only be developed after many long, unwashed days in damp bothies. In my experience, such badge-wearing bothy visitors are, without exception, nutters. Not that I mind nutters – I've met quite a few in bothies, and there may be a sizeable portion of the population who consider me one. This particular brand of nutter is always pretty tedious at best. They start out quite friendly and seem reasonable enough, but then as they make your acquaintance, late in the evening, they always begin to unveil their particular type of conspiracy theory. This frequently covers every type of fantastic scenario from the Chinese invading us by tunnelling under the Channel to the Mountain Bothies Association being an undercover plot to make money from the commercialisation of the use of bothies. Their delusions are always ridiculous, but they are always unshakably convinced of the validity and righteousness of their beliefs. Rational argument is a waste of time, for theirs is an obsession. It's what defines them – take it away and they collapse on the floor like a deflated sex doll. They, of course, can never see that they are obsessed, because if you have the perspective to understand that you're deluded, then you're not obsessed.

'No thanks, no tea. Better be off, I think.' Then I remember the sack of coal in my rucksack.

The poor little man looks crestfallen. 'But, but, but … ' is all he can say.

'Would you like some coal?' I ask to try and brighten his day.

The badge-wearer couldn't look more delighted if I'd offered him a gold bar. 'If you don't mind. It's been freezing.' There's something about the way he says the word freezing that tells me he has spent long nights shivering.

I haul the bag of coal out and place it on the grass between us. His delight redoubles when I also set down some kindling and firelighters. At that moment, if it wasn't for my expressed concern about the virus, I am pretty certain he'd have kissed me. He would probably prefer coal to gold right now, as you can't get warm by burning gold.

He is effusive in his thanks and rushes into the bothy with his prizes as I hoist my rucksack on to my back and begin to walk away. By the time I am almost back at my car, there is a plume of dark smoke rising from the bothy chimney. The badge-wearer might be alone, but at least he'll have the fire for company.

I drive away with the sinking feeling that my bothy days may be over for a long time. More than that, if I am banned from travelling, the wild winter that I have planned to diarise will be cut short. On my way back home, I have the uneasy sensation that I am driving into the unknown. Perhaps bothy nights and books to write will be the least of my problems.

***

I've only been back in my flat in Inverness for an hour when the phone rings.

'John, it's James. There have been marten at the hide for the last three nights at around 7 p.m. Do you want to come and see them?'

I can't contain my excitement. 'Yes, I'd love to.'

'Okay. Tomorrow night. I'll meet you at the hide.'

James's hide is on the Black Isle near Inverness. Just to clarify things for you, the Black Isle is neither an island nor black. James lives near the village of Avoch, which is pronounced 'Och'.

Is that clear enough? This was designed to confuse the English. It certainly confuses me.

When I arrive at the lay-by beside the dense woodland, James is leaning against his car waiting for me. He leads me off into the woods, and after a few minutes we are standing in a clearing. The hide, which is basically a garden shed with half a wall removed, overlooks an angled log. This log is the feeding station that James uses to lure animals into the open so that his photography clients can take pictures. I have an ever-growing admiration for wildlife photographers. So far this winter, I have roamed about trying to find otters and mountain hares. It's been a long, cold and, at times, unrewarding business. Add to that the fact that I am merely trying to see wild creatures and not capture images of them, and you begin to realise what a hard game wildlife photography is.

James sets a couple of eggs and some peanuts and honey on the log. 'I've been doing this for two years now. It's taken me that long to get their trust.'

He moves about the woodland with the quick agility of someone used to moving in wild places.

'Do they always come?' I ask.

'They are pretty reliable,' he whispers excitedly, 'but sometimes they get waylaid, maybe find some unexpected prey or something scares them off.'

James always talks in a quiet hush. All his years seeking wild animals have imbued his speech and even his actions with a measured quiet. I can't imagine him shouting, even if he wanted to.

He finishes by covering the peanuts in honey. 'Usually they

follow a pattern. Recently they have been coming at about seven.' He turns to head into the forest and back to his car. 'Remember, keep still and try not to make any noise.'

James leaves me to watch over the baited log. I settle down. It's still winter, but the days have lengthened now. The short, dark days of December are behind us and spring is creeping in at the edges. A few weeks ago the sun was pale and weak, but now it has some warmth in its rays. After the cold, stormy months, life is coming back. It returns to the land just as surely and irresistibly as the tide returns to cover a sandy beach. There are buds in the trees around me. Soon it will be April, when the land will stir, and then the riot of May will arrive. May is one of the great months in the Highlands. The days are long and the air fills with the buzzing of insects as life explodes across the hills. Not long to wait now. I love the winter, but by the end of March even I will be waiting for the call of summer.

I sit and wait, watching as the light begins to fade and the forest sinks into darkness. Only James's illuminated log stands out against the deep dark of the forest. It reminds me of a film set. The artificial light catches the log and a few bushes, making them too appear fake. The cold is beginning to creep into me now. My upper body is warm, insulated by my down jacket, but my legs, despite my hill trousers and an extra layer of long johns, are chilling rapidly. I desperately want to fidget, to move and get some blood back into my legs, but I know I can't. Despite the cold, I must remain still. I'm almost literally frozen.

My closest encounter with pine martens took place in November last year. I was sleeping in the bothy on the Ardtornish estate. Around three in the morning, I became aware of an odd noise. It seemed a long way off, so I decided to ignore it. I pushed it to the back of my mind, snuggled into my sleeping bag and began to drift back into a downy sleep. The noise came again, a sort of scratching, tearing sound. This time it was closer and more persistent. In the darkness of the bothy, I tried to clear the sleep fog in my brain. There was something on the roof. I sprang into action. I planned to grab my head torch and catch whatever it was. What actually happened was, I fell off my inflatable mattress and rolled around on the floor desperately trying to extricate myself from my sleeping bag and locate my head torch. The flailing around lasted about ten minutes or twenty seconds, depending on whether you use real time or the impressions formed by my panicking brain.

'Piss off!' I yelled.

Whatever was on the roof ignored me and the noise persisted. That alarmed me even more. There was a wild animal trying to get in to the bothy and it had ignored me. If it was not scared of me, then I was scared of it. At last I wriggled free of the sleeping bag and my fingers found the metal of the torch. I got up off the floor and switched on the torch. Light! I was vertical and no longer blind. The animal on the roof must have heard me, yet still it was trying to get in. I lurched towards the sound and shone the torch upwards. The noise stopped abruptly and I could hear something moving quickly across the slate roof. I half fell, half ran downstairs and rushed out into the night, searching the tall grass around the bothy. I wasn't sure what I expected to see as I blundered down the bothy stairs in my underwear. Outside the bothy, a bright moon highlighted the clouds scudding across the night sky. My head torch beam picked up the tall grass waving in the wind, but that was the only movement. There were no bright eye spots watching me from the undergrowth. Whatever was on the roof was long gone.

That was my first close encounter with a pine marten, the only creature that could have been attempting a burglary at the bothy. In some areas of the Highlands, like the Cairngorm National Park, pine martens are relatively common, and some like to live in the roofs of houses. That probably explains why my night-time visitor was trying to get in to the bothy roof. Despite their relatively healthy populations in some areas, they are rarely seen. Pine martens are creatures of the forest and are nocturnal, so their wanderings are rarely observed. They will emerge from the woodland to hunt, but are naturally wary of humans. In all my time in the Highlands, I've only caught fleeting glimpses of these creatures, mainly in my car headlights when I have been driving home at night from the hills and they have been crossing the road. On the tarmac, they look odd and ungainly. Pine martens are built for a life in the trees, where they mostly hunt squirrels, young birds and eggs. I can only hope I will see one tonight.

Forty-five minutes later, I am still sitting motionless. It is not totally dark yet and my eyes have adjusted to the gloom. In the half-light, I begin to see movement everywhere. My mind tricks me into thinking that every bush, every shadow is a creature. I start to hear things. At first the forest seemed silent, but now I hear rustlings and creaks in the trees. There is no wind, but I have the distinct impression something is moving behind me. I think I'm getting hide fever. The log and the bushes are motionless when flakes of snow begin to drift through the light beam illuminating the log. They sparkle white before falling to earth. It's only a light flurry but it adds a sense of wonder to the scene before me.

Then it happens. Just off to the right, I see the bobbing back of an animal. There is a pine marten heading for the log. I wait, hardly daring to breathe. At first she reaches up to the torch James set up and sniffs at it – perhaps she can catch the human scent of his last touch. Then she climbs nimbly up on to the log. There is beauty in the way she moves so quickly and with such agility. She has the grace of a ballet dancer with the speed and precision of a deadly predator. Her brown body is lithe and her movements liquid as she approaches the eggs. Then she freezes and, turning towards the hide, fixes me with her eyes. Her stare seems to look right into my core. This is not a casual glance. In her eyes, I can see the wild; an ancient instinct focuses precisely on me. She must decide in an instant if I am a threat or if it is safe for her to linger. The wrong decision could cost her life. People are her enemy, yet she has learnt to tolerate their presence, to walk into a beam of light that could bring her death just as easily as food. Her eyes are fixed upon me while she makes a decision.

I have seen that look before. Once I walked into a clearing in a snowy Canadian forest at the exact moment a mountain lion entered the same clearing. The lion froze, just as the pine marten has done now, and stared into my eyes. The look only lasted an instant, perhaps half a second, but in that time both our fates were decided. At that moment, the lion was deciding if I was a threat or if I was prey. Fortunately for me, she decided I could be ignored and casually walked away, just carrying on her business in the forest. As I watched her rippling body depart slowly, I realised she could have attacked me. Oddly, I felt no fear but was filled with a kind of inner calm. A moment later, the clearing was quiet again, the only sound the beating of my heart.

I know that if I move while the pine marten examines me she will be gone in a flash of fur. After a few seconds' scrutiny, she decides I am no threat. Obviously she has experienced people in James's hide before, so to her I am just one of those odd humans who sit in the box near the eggs. Calmly, she takes one egg from the log and scampers off, returning moments later for the second. James had told me she would do this. An egg for a pine marten is a good protein-packed meal, one even worth taking the chance of being seen for.

I wait a few more minutes and she returns. She's had her main course; now she's ready for a dessert of peanuts and honey. The sticky nuts are more difficult to access. She can't just grab these and run. For several minutes, I watch her grabbing nuts from the feeding tray with quick dexterous movements. I'm tingling with excitement; to be just metres from such a creature is a precious gift from the natural world. I find it difficult to believe that for the first time in all my years in the Highlands, a wild marten is only a few feet from me. This is no zoo. I am not watching some bored caged animal through wire mesh whose behaviour has been irradicably changed by years of psychotic captivity. This animal was born in the wild, has faced and overcome danger. Here before me is that moment of wildness I sought. I no longer feel the cold – only joy at being so close to such a wonderful creature.

The pine marten finishes her dessert, gives me a casual glance and exits, like an actor from the stage of her log. I wait five minutes, hoping she'll return, but now the peanuts and eggs are gone I decide she is unlikely to come back tonight. My knees are stiff with sitting for so long in the cold as I pack up my flask and rucksack. To my amazement, just as I am leaving, the pine marten returns. She is startled to see me outside the hide and bounds off towards the forest, where she climbs a pine tree with incredible agility. The last I see of her is her head popping round the back of the tree to take a look at me. It was an accident, but part of me feels I betrayed her trust. When I check the tree, I find she's only eaten about half of the nuts. I'm sure she'll wait until I've gone and come back, moving with ghost-like silence, to enjoy the remainder of the nuts. I set up my camera trap to watch over the log. I'll leave it here for a few days to capture the visits to this secret place. The pine marten will never know she's become a film star. I walk back to my car through the silence of the forest, knowing I have achieved something special. I have made a connection with a wild creature, and the memory of those few minutes in the secret world of the pine marten will stay with me for the rest of my days.

WALKING INTO SUMMER

When I look up from my computer screen, which is often as I am struggling to find the words to type, my eyes are drawn to the branches of a sycamore tree only yards from my window. I am a few floors up so I don't see the trunk of the tree, only its uppermost veined branches as they dwindle into the sky. I know that the tree has always been here, but when I first moved to this flat its branches did not reach my windowsill, and so I looked out over it to the rooftops and the wooded hills beyond.

I am trying to conjure words about my memories of journeys to wild places, but find myself sitting for minutes on end suffering from some strange paralysis. My fingers hover over the keyboard, words remain unconsolidated in my mind and I find myself studying the intricacies of the thin branches against the grey sky. To make matters worse, I am constantly distracted by a tiny comedian. A blue tit has become a constant visitor to the sycamore tree, and he delights in catching my eye every time I settle down to type a few sentences. He seems to sense when some kind of flow has begun to develop in my work and chooses that moment to appear at the window. I admit I have encouraged him. I noticed him in the branches a few weeks ago and enjoyed watching him so much that I wanted to entice him closer, so I purchased a round plastic bird feeder and stuck it outside my window.

There are other visitors to the sycamore: bullfinches, greenfinches and a little flock of sparrows. There was even a kestrel sitting there one morning when I sat down to work a couple of days ago. Only the blue tit is cheeky enough or bold enough to risk visiting the feeder just outside the window. At first, when I looked up, he would flee in an explosion of yellow and blue feathers. Now he has got to know me. When I look at him, he watches me with quick head twitches and regards me with mild disinterest before resuming eating the balls of fat in the feeder.

We are three weeks into lockdown and the world has changed with a devastating suddenness. The days I write about – travelling the Highlands visiting bothies, hills and lochs – are only a memory, dreams of another world. Was it me who watched the pine marten stealing eggs, or lay in the snow just feet from the dozing mountain hare? Was I once that person, and do such places still exist? The sycamore tree and the birds who occupy it are my only contact with the natural world. You would think that for a writer, being confined to barracks would be an asset. Perhaps with no distractions I would become super-productive and hurl myself into writing. I tried to do that, telling myself that three weeks is not a long time to be away from the outdoors and I would soon be back and I should get on with things. Yet every time I sit down at the keyboard, a strange lethargy overtakes me, and it becomes harder to focus with each day that passes. At first I could not understand why this was. I could go out for exercise once a day for an hour, so I was getting out. These walks felt alien to me, as if I were a prison inmate, allowed to walk each day around a yard. I have never been an urban walker. I couldn't see the point of walking the streets. I began exercising at home, doing press-ups and sit-ups and squats holding heavy objects to make the exercise more effective. I think I managed about a week before I could no longer motivate myself. All across Britain and the rest of the world, there were people far worse off than I was. I told myself I should be happy I was healthy; it just didn't feel that way.

Now a sense of unease undermines everything I try to do. On Twitter, people talk about how they have super-cleaned their homes and done myriads of DIY jobs they had been putting off for years. In contrast, I let the dust build up, leave dishes and tea cups to fester. I go unwashed and rarely change my clothes. I am certain I stink. I catch my body odour every now and again. This doesn't seem to matter to me anymore. I live alone and, as I'm certainly not going to get any visitors, being smelly doesn't matter. There is no one to notice the dishes in the sink or run their finger over the dust on my TV set, so why bother? I am usually a self-motivated person. I am always doing something, eager to be out in the hills or writing or creating something, but instead I find myself wandering aimlessly around my flat. If I notice something that needs doing, I simply put it off. After all, I have all the time in the world.

I'm sitting staring at a blank page on my computer screen when the phone rings and I hear Martin's familiar voice. 'How are things with you?'

He sounds flat; neither of us can find much positive to talk about.

'Have you been anywhere?' I ask, realising as soon as I speak how stupid that question must sound.

'No, nowhere. I've resigned myself to not going anywhere for the next two years.' He sounds defeated.

'You'll still be able to use the trains, won't you, when things ease a bit?' I try to encourage him.

'No, I don't think so. Any trains are going to be very limited. I can't stand these mask things. I won't wear them.'

Martin has travelled by rail all over Britain his whole life – it's been his passion. Without it, he'll be a prisoner, as he doesn't drive.

This means the end of our hill days together. 'So, no more trips to Scotland, then.'

Martin sighs. 'All that's finished.'

I try to think of something to say that will give him hope, but can't find anything. By the end of our call, I am beginning to realise that the times I have spent together with my old schoolfriend might be confined to the pages of the books I write. That's a painful conclusion for both of us.

Yesterday I found myself staring out of the window. I realised I had been sitting for almost an hour watching the blue tit and his mates flitting about in the sycamore tree. At other times, I've gazed out across the River Ness as it runs past my house untroubled by the pandemic. At least some things are still normal. I found myself touching the window glass, pushing against it gently at first and then harder. Then I realised why I felt how I did, why my energy was gone: these windows had become the boundary around my life. My flat was now my safe zone. Outside, just beyond the glass, a deadly miasma lingered.

All my life, I have found release in open spaces, the boundless world of forests and hills. I have always known that when the pressures of life overwhelmed me, I had somewhere to run to. When I was unemployed and couldn't find work, or when I was working and the stress got too much, I knew where to go. When my father died and I couldn't accept he was no longer in my world, I knew that what I needed was to find somewhere wild, green and free. After months of visiting him in hospital, walking along sterile corridors, I eventually watched him slip away. It was then that my own health collapsed. I had been keeping going, not realising the toll it had taken. Still weak and exhausted, I went to Sutherland. There, on the wild north coast of the world, with the wind and waves around me, my spirit healed.

My problem now is that my world has been reversed. The places that had always been my safe haven are occupied by the enemy. The dangerous place to be is not this side of the glass but outside it. I have nowhere to go. Without my safety valve, I am lost. For a few days, I sit feeling helpless, my world shrunk to the four walls of my flat. I watch the birds in the tree and look out over the river, but I can't find any drive. Watching the daily bulletins on TV only adds to my sense of being under siege. It is depressing watching Boris pretending he is in control and everything will be fine when so clearly he isn't and it won't. It becomes clear to me that even if I can't go to the wild places, I can't stay in my flat for days on end. The problem is that my flat is situated in the centre of Inverness. The Highland capital works well for me as the transport hub of the Highlands. In almost any direction, an hour or so's car journey will take you to the hills and away from urban life. That only applies, of course, when you are allowed to make such journeys.

***

At first, the fact that I can only walk in the urban town centre doesn't inspire me, but if that is all I can do, then I'll have to just get on with it. I decide I will walk every day. On my first walk, I set off to follow the River Ness upstream towards the town centre. The roads are eerily quiet. The bridges over the river usually bustle with cars and lorries, but now only one or two vehicles pass by. There are few people about. Those I meet are clearly escapees from lockdown just like me. We pass awkwardly, keeping our distance, unable to disguise the fact that we are afraid of each other. The river babbles past on my left as it always has, but the road on my right is surreal and still. It's as if I'm in some zombie apocalypse movie, frightened that the people I meet might be on the change. There are ducks sleeping in the middle of the road, exuding the attitude that since the human race is evidently on the way out, they'll just take over. I walk past the shuttered hotels and shops with makeshift signs that make it clear that they are closed and have no idea if they'll ever open again. I go as far as the red brick cathedral, its doors gated and locked. On the far side of the river, the Victorian castle, now law courts, looms over me. Usually the whole place is thronged with locals and tourists; today there's me and one middle-aged woman in a tracksuit who eyes me suspiciously and crosses over to the other side of the road.

In the quiet, I sense a pervading atmosphere of fear. I am taking all the precautions I can think of. I don't think I am any more frightened of dying than anyone else, but I am on strict instructions from my daughter, who is a frontline worker, to take no risks. Besides, I am due to draw my old-age pension next year and I would feel particularly aggrieved if I was to shuffle off this mortal coil before the state was due to dole me out cash. Despite my initial reluctance, after a week or so I begin to look forward to my urban walks. There is something inherently natural in the sensation of walking. It's something the body is designed to do and, despite myself, I find my daily walks begin to lift my mood. I look forward to them rather than treating them as something I should do because they are good for me.

It's early April now and there are signs, even in my urban environment of pavements and tarmac, that spring is on its way. The sycamore tree is still leafless, but its outermost branches are tipped with buds that seem on the point of bursting out. Writing is still difficult, but at least now when I look into the tree there is much to see. The blue tit is making daily raids to my bird feeder and there are a host of other birds constantly feeding on the buds of the sycamore. I'm not sure what attracts them – perhaps there are insects there – but they are busily gorging themselves on something. I watch a small flock of sparrows as they rampage through the branches like a gang of schoolboys stealing apples from an orchard. The blue tit is often joined by two or three bullfinches. I had seen the finches before, but never realised how colourful they are. Because my window is at the same level as the top of the tree, I can focus my binoculars directly on the birds and carefully examine their plumage. Another greenish-yellow bird appears that my little book tells me is a greenfinch.

The buds opening on the sycamore indicate that the seasons are changing. One morning, as I am walking past a silent hotel, a mallard emerges from the river followed by around a dozen ducklings that can't be more than one or two days old. They follow her, in a brown chirping procession, over the pavement and on to the road, which is devoid of traffic. Once over the road, she heads to one of the upmarket hotels, where they vanish under a bush which she has made her home. I can't help smiling at this natural event right in the heart of the town. It makes me realise that the winter I was writing about when Covid-19 arrived has all but ended. Spring is bursting forth everywhere in the hills, but I am missing it for the first time in my life. This year, I didn't see the pools on the hill tracks bursting with tadpoles. I haven't heard the cuckoo calling across a quiet glen. By now, the hills must have lost their winter hue and be turning green as life awakes from its winter slumber. It feels odd that I am not there to see these events that I have always looked forward to because they mark the changing of the natural calendar.

The lockdown weeks are passing. The number of deaths climbs each day and there seems to be no end in sight. One dull April day, I am on my daily ritual of a walk when I see a small black bird skimming the waters of the river. It's a house martin, the first I have seen this year. I stand watching its acrobatics as it hunts for insects a few inches above the surface of the water. This little bird has just flown all the way from Africa – a miracle of flight and survival, thousands of miles of travel across seas and continents to arrive here on the river in Inverness. If that fragile creature can do this, then I wonder what we can achieve. In these dark times for me, there is a message of hope in the arrival of the first house martin. The next day, there are four or five martins dancing over the water. Two weeks later, I am stopped in my tracks staring at thousands of these small agile birds flying back and forth. I have never seen so many flying endlessly across this small stretch of water. They are joined by swifts cutting endlessly back and forth over the Ness, hunting for insects. I walk for over a mile down the river, and for that whole distance the air is full of the black shapes of these tiny fighter planes performing split-second manoeuvres. Perhaps the lack of air pollution caused by lower car use has caused the insect population to boom. The myriad swifts and martins dancing in the air above the river are a sign for me that, despite all that besets us, the heart of nature still beats.

I notice finches and tits in the narrow strip of bank between the river and the road. This area, only a few metres wide, used to be mowed regularly by the local council. Now they seem to be letting it grow wild and nature is claiming it back, as it always does when given the slightest chance. I'm not sure if the lazy ducks still sleeping in the empty roads or the finches chattering in the tall grass beside the river were always here and I am only noticing them now, or if the lack of traffic and mowing has helped the environment recover.

As the days and weeks pass, I begin to enjoy and extend my walks along the river. At last, we are allowed to travel five miles from our homes for exercise. Unfortunately, this does not help a great deal, as when I drive five miles I discover that everyone else in Inverness has had the same idea. Everywhere five miles from my home is crowded with families out walking or folk with their dogs. I'm glad to see them out, but it means that the solitude I so enjoy is hard to find. There is beginning to be talk of our countryside opening up and I am desperate to visit the mountain bothies I love so much. The problem is that they are the perfect storm in terms of transmitting the virus. If you visit a bothy, you have no idea who has been there recently and who you might be sharing the bothy with that night. If there are folk there, then they may not be good at social distancing; and even if they are, most bothies are fairly small and make that difficult. Bothies don't have plentiful supplies of hot water, and are often places where dust has accumulated over many years, so cleaning them effectively is impossible. The virus is also said to thrive in cold, damp environments, which is exactly what all bothies are. Put simply, if you want to spread Covid-19, then a bothy is an ideal place to do it.

Perhaps by the time you read this book the fate of the mountain bothy will be clear, or possibly there will be a vaccine and everyone will have carried on as before. That's something my readers will know and I cannot. Only a couple of years ago, I wrote my book *Bothy Tales*. That book did exactly what it said on the tin: it contained stories about bothies. What I didn't know when I wrote the book was that perhaps I was writing about a culture that was destined for extinction. It seems hard to imagine in these days of the pandemic that there was once a time when you could wander into a bothy, meet with a group of like-minded people and spend an evening before the fire without a care in the world. Perhaps that world has gone forever and my book will one day be confined to the shelves of an academic library, where a sociologist will read it as an example of the changes wrought on the culture of the early twenty-first century by the coronavirus plague. I sincerely hope that isn't the case, because if that happens we will have lost an important part of our mountain culture. Bothies are one of the few places you are able to meet folk as diverse as a Glaswegian bus conductor, a Dutch eco-warrior and an ageing writer. Such meetings allow us to forge bonds, to find the commonality between us all. It is the wild surroundings of bothies that make them unique. Nowhere else do we realise that we are all frail humans in the face of nature. They are a long way from the sanitised hotels we stay in or the noisy commercial bars people share. They are real places, and I hope they remain so.

In the short term, until we find out if bothies are to be consigned to history, I decide to find an alternative. I need to get a tent that will allow me to replicate, as far as is possible, the bothy experience. I have never been a great fan of camping. I have done a lot of it over the years, but the idea of spending a holiday voluntarily living like a homeless person has never appealed. I have camped in a great many places, not because I enjoyed camping but because I enjoyed what camping allowed me to do. When I was young, I couldn't afford any other accommodation. Bothies have allowed me to travel the Highlands without the need to crawl on my hands and knees through a nylon obstacle course to go outside for a pee or to perform some yoga-like contortion in a desperate struggle to put my trousers on. I admit that many younger folk will find such manoeuvres considerably easier than I do. I, however, am a lot less flexible than I was in my twenties, and also a fair bit bulkier.

I need a tent I can stand up in, even if only for the purpose of trouser application. I also want a tent I can heat fairly easily in order to carry on my adventures into the winter. As far as heating the tent is concerned, I hit on the idea of buying myself a tilley lamp to provide my substitute bothy. For those of you less ancient than me, tilley lamps are pressurised paraffin lamps that provide both light and heat. They are, of course, very heavy, but I have already decided that I don't intend to carry my portable bothy more than a mile or so from the road, so weight is not a major consideration. Tilley lamps have now been confined to museums, but they are still available on places like eBay, and there are whole squads of men in sheds who lovingly preserve them and even collect them. I know that there are perfectly serviceable gas heaters and lights for tents, but somehow these seem too modern for my portable bothy. I want something like me, a little eccentric but without going full badge-wearing survivalist geek.

I come upon the ideal tent on Amazon, an odd-looking tepee design made in China. It is six kilograms and offers me easily enough headroom to allow me to get dressed without slipping a disc. I promptly order it, and at the same time put some bids in on eBay for a tilley lamp. To my surprise, I am quickly outbid for the first few lamps I try to get. They seem to be in great demand. Perhaps the number of survivalist geeks is growing.

The pandemic has vindicated all the survivalists in the world. Right now, they are sitting patting their stocks of three years' worth of toilet rolls, smiling to themselves. 'Well, they aren't laughing at me now, are they?'

My lamp duly arrives and, after a few mishaps involving columns of flame shooting up towards the ceiling, I manage to get it purring away nicely, illuminating my back bedroom and filling it with warm if odd-smelling vapours. I am slowly preparing for post-lockdown, although right now it doesn't matter what tent or magic lanterns I have, nobody is going anywhere. There are rumours things might ease in July. The whole outdoor community in Scotland is holding its breath.

***

Despite the pandemic seeming to dominate, everything ordinary is somehow carrying on. In June, something extraordinary and incredibly significant happens. Completely out of the blue, the Scottish parliament introduces legislation to make mountain hares a protected species. This move will end the mass slaughter of these wonderful creatures, something that has been going on for over a hundred years. It is estimated that the mountain hare population is now only one per cent of what it was seventy years ago, and around 26,000 have been slaughtered every year, almost exclusively on moors used for driven grouse shooting. Shooting estates had claimed that the hares spread the louping-ill virus by increasing the number of ticks. These assertions have little scientific support, but the blood sports community never worry too much about science, as the men in tweed's response to most problems is to kill something.

Unsurprisingly, sporting estates react to the news with horror, predicting the doom of most rural communities as a result. Alex Hogg, chairman of the Scottish Gamekeepers Association, says the Holyrood vote is 'a grave mistake. This is a bad law, made by people it will not impact upon. The views of the rural working people of the land have been ignored here. The system has failed them.'

The Gamekeepers Association uses one of its tried and tested public-relations strategies of asserting that those who wish to impose controls on blood sports practices are people who don't live in the countryside. In other words, they are trying to promote the myth that those wishing to protect wildlife are 'townies' who have no real understanding of life in rural areas or the wildlife they are trying to protect. In contrast, they portray the men in tweed as 'guardians' of the countryside whose only wish is to protect the natural environment, despite the fact that under their stewardship much of the Scottish Highlands has become little more than a green desert.

The association that seeks to protect the interests of landowners also makes a statement prophesying doom. Sarah-Jane Laing, chief executive of Scottish Land & Estates, says, 'These changes will not help Scotland's wildlife, which is the prime concern of gamekeepers and land managers.'

In other words, the best way to protect mountain hares is to shoot them. This legislation, which is passed alongside a law to prevent salmon farmers from shooting seals, has far greater significance than extending the protection of the law to mountain hares, important though that is. It signifies a profound shift in the balance of power in the wild areas of Scotland. In many ways, the Highlands have been frozen in time for as long as I can remember, with large estates and their owners having almost complete autonomy over what happens there. Land is power. In Scotland, land ownership is concentrated in the hands of fewer people than anywhere else in Europe. Only around 400 folk own about fifty per cent of Scotland's private land. That represents a huge area of land in the hands of a tiny fraction of the population. The owners can pretty much do what they want with it; much of the legislation aimed at controlling what they do is either toothless or has been blocked by the political power of wealthy landowners. A good example is the 2019 attempt to control hill tracks. In recent years, the hills of the Highlands have been increasingly scarred by tracks bulldozed across them.

Some of these are for the construction of hydroelectric schemes, but many are simply for the convenience of sporting estates who just want to be able to get around their estates more easily. Once created, these tracks are more or less permanent and the damage is irreversible. Attempts were made to introduce tighter planning controls for these tracks, but they were blocked in the Scottish parliament by the interests of landowners.

It always seems incredible to me that what happens to vast areas of land is still at the whim of the owners. As land reform campaigner Andy Whitehead points out, no one can really *own* land – all they can have is rights over that land. Land isn't like a car or a TV set. Hundreds of years ago, no one owned the land until someone established rights over it, usually by violence. If you could hold on to the land by beating or killing anyone who tried to take it from you, it became yours. Here we are, in the twenty-first century, still living our lives in a way governed by punch-ups hundreds of years ago. Is land not the most important commodity of all? Without it, who are we? What are we? The land is what we come from, and yet this absurd system remains and our land is still controlled as in feudal times.

I have often wondered why those who own big estates fight so hard to preserve them as they are; as far as I can see, it has little to do with making money. Highland estates are not owned to make profit. They are owned in the way a billionaire might buy a football team. Perhaps it's for the lifestyle or just the sheer pleasure of knowing that you own a chunk of Scotland. There may be a certain sense of tradition about what is clearly a way of life with its rituals and customs. I suppose it's nice to own a castle, although I wouldn't like to pay the heating bills. These things may be factors, but I think the reason that the powerful men who own Highland estates fight so hard to do as they wish there is far darker than that. Ultimately, the determination of landowners to continue to follow pursuits like driven grouse shooting is about power. It is something deeply embedded in the class system of our nation. It's a way of showing working-class folk that this is the owners' land and they can do what they want with it.

There are castles and grand houses scattered all across Scotland. In some ways, they are fascinating survivors of hundreds of years of history. But take a closer look. Most castles across Scotland are crammed with one thing: the symbols and tools of death. You will find suits of armour and ancient swords, spears and axes lining the walls. To us, these are interesting artefacts, curiosities that fire our imagination. When they were manufactured, they were something very different: they were state-of-the-art tools of oppression. The weapons that we view with academic interest were made to inspire terror. The big stately houses of Scotland should also exhibit the injuries these weapons were capable of inflicting. Instead, they content themselves with celebrating the death of wildlife with rows of stags' heads to demonstrate just how efficiently they can kill things. I suppose even the bloodthirsty ancestral owners of Scotland's sporting estates realise that it would be a bit off to display the corpses of peasants. Most estates were acquired through violence by their original owners, and they were held on to by the threat of violence. The plush interiors that we see were assembled while large swathes of the population lived in poverty. I can never forget that these places were founded upon oppression.

In Scotland we have the right to roam across any land, a concession that I am sure rankles with many landowners. The signs that you can see at the foot of some hill tracks warning you that shooting is in progress and it's dangerous to proceed when it clearly isn't are evidence of that. We are allowed to walk on the land, but not to have any say over what happens on it. The prohibition on the slaughter of mountain hares that was passed by the Scottish parliament is not only a significant step forward for the protection of wildlife, but also immensely important as an indication that the landowners' control over what happens in our countryside is beginning to crumble.

Signs that the days of the traditional Scottish sporting estate are numbered are also visible in the economic value of the land. In recent years, when estates have come on to the market, interest in those with traditional sporting assets has waned, while land with the potential for rewilding and carbon sequestration has seen an increase in interest. This reflects wider changes as environmental issues move higher up the political agenda. Obviously, you need deep pockets to be able to purchase your own Scottish manor. In 2018, the value of Scottish estate sales totalled 86 million pounds. Most of us would need to save our pennies for a few hundred years to be able to buy our very own turreted mansion. For a long time, these places were literally the playgrounds of the landed gentry, but in recent years the weakness of the pound against the dollar has led to increased interest from America, and the environmental potential of some estates is attracting a different sort of buyer. There are now financial incentives in taking grouse moors and turning them into forests that can sequester carbon, so a whole range of new opportunities for the use of large areas of Scottish wild land is opening up.

There is a growing environmental awareness amongst a large section of the population, and it may well be that the number of people who want to spend their leisure time shooting a stag or blowing grouse out of the sky is diminishing. Put simply, the shooting culture, even amongst our landed elite, is literally dying out. Many estates are beginning to diversify. The Ardtornish estate on the west coast of Scotland is a good example. The estate invested 15 million pounds in a hydroelectric scheme and offers accommodation and fishing as well as being a wedding venue.

As Hugh Raven, managing director of the estate, says, 'We needed to try to find long-term sources of revenue because landed businesses are pretty marginal and quite expensive to operate.' Their income generated from hydroelectric power is 'twice as big as all our other sources of income put together.'

Change is coming to the wild areas of Scotland, and it is being driven as much by economics as by ideology. There is a real opportunity for Scotland to become a much more diverse and vibrant land, a place with a variety of habitats and its rural economy funded more by the binoculars of wildlife tourists than the guns of blood sports enthusiasts. There is hope for the return of life to the glens and an end to the tyranny of death at last.

***

It's more than three months since Boris stood at his lectern in Downing Street and switched off my winter. I'm in a petrol station for the first time since he made that fateful speech. For the last three months, my car has been reduced to supermarket trips, with its longest journey being an eight-mile run to take some food to a friend who was in isolation. Now I am travelling to Achnasheen and I will be the first person to walk into Inver hut since lockdown was imposed. The boot of my car is full of bottles of handwash, surface cleaner and disposable cloths. Bothies are still closed, but the Scottish government has allowed huts owned by mountaineering clubs to open under very strict limitations. As I am only fifty miles from the place, the Jacobites Club, who are Edinburgh-based, have asked if I'll go and see how their hut has fared through the weeks of solitude. This is, of course, something I am delighted to do, as it offers me the prospect of a few nights in the hills, a thing I have been longing for.

I head out of the town and over the bridge across the Moray Firth for the first time in months. I haven't been over thirty miles an hour for weeks and I have almost forgotten how to drive. The hedgerows flash by at incredible speed and the tarmac rolls beneath me so fast I grip the wheel tight, hands sweating. I risk a quick glance at the speedo. I am doing fifty-five miles an hour. It is surprising how quickly you can lose a skill that seems so ingrained. It takes me twenty minutes of driving before I feel comfortable at the wheel. After forty minutes, I am driving into the hills for the first time in months. I am staggered by how much the landscape has changed since I last saw it. There has been a green explosion. There is a riot of burgeoning life from the trees at the roadside, heavy with new growth, to the hillsides above, verdant with lush grass. Summer has arrived, but I have never seen it like this before.

The last time I was in these hills, it was in March when winter held sway. Then the grass was yellow and lifeless, the heather dead and brown, the trees leafless. I had forgotten that while I was imprisoned in my home in Inverness, summer was stealing into the landscape. It is the sudden contrast between the winter hills I left back then and the summer mountains I see now that astounds me. Of course, this change has happened every year; the only difference is that in the past I watched this change take place gradually. I saw the trees come into bud and the grass start to grow. I saw the insect life gradually come back to the air and listened as birdsong returned to the wind. This year, I saw none of this and missed spring entirely. It's like when you remember a young boy of ten singing and playing with boats and then see him again when he is eighteen. In your mind, he is still playing with boats, but when you meet him again he smokes roll-ups and towers over you.

I arrive at the lay-by where the path to Inver leaves the road, crosses a small bridge and then passes over the walkway across the bog to the stone cottage a few hundred metres away. There is a stile over the fence from the lay-by to the bridge that is normally obvious, but now I see no sign of it. I find it buried in bracken, the path overgrown and choked with grass. It reminds me of a post-apocalypse movie when the hero awakes from a coma to find he is completely alone and nature is reclaiming the man-made world. I guess Covid-19 has been our apocalypse. It's an eerie feeling pushing the door of Inver open and knowing that I am the first person to step inside for over three months. The place has a dank smell, as you might expect from a basic cottage that has had no ventilation for all this time. The lounge where the fireplace sits is covered in a fine layer of dust, perhaps ash that has blown down the chimney over the months. I throw open the doors and windows to let the air circulate for the first time since March. I want to let the place breathe.

I too need the air. I walk out of the hut and stand revelling in being out in the wind once again with the smell of the wild and the calls of wild creatures. It feels as though I have woken from a nightmare. Yet there are reminders that the disaster of the virus still stalks us. It is July and I would have expected the road near the hut to be busy with tourists, locals and passing trucks, but it is quieter today than it would normally be in December. I'm just about to return to the hut when a pair of sea eagles swing gently over the ridge behind the hunt. Two huge shapes in the sky, their size makes them unmistakable. They soar along the shore of the loch and then turn away from the water to head over to Strathconon, perhaps searching for carrion. I see them for less than thirty seconds, their great wings beating so slowly yet covering enormous distances with an effortless grace. I haven't seen sea eagles since my trip to Mull, and I'm so lucky to catch sight of them on my first trip after lockdown.

In the evening, I sit in the hut watching the flames of the first fire Inver has seen for months, feeling like a prisoner released from jail. I am on the verge of recapturing the life I once had. I remember lying on the clifftop watching the seal pups in Sutherland in November, and my joy at finally capturing a close-up view of a pine marten in the cold of a March night. How long ago that seems now, how much has changed.

I lean forward to put another log on the fire and catch a movement outside the bothy window. It's a red deer, a hind, tiptoeing past the window in the fading light. Another hind follows her, only feet from the window of the hut. Perhaps Inver has been empty for so long the deer have forgotten it was often occupied and have lost their fear of the place. Two more hinds pass by, and this time each of them is followed by delicate fawns in the spotted coats designed to camouflage them in their first few days of life. The fawns are only a week old and stay close to their mothers through the long grass, on thin legs that look as if they can barely hold them. I began this book in October, when these young creatures were conceived. I had planned to end the book by seeking out the fawns who were born following the rut I witnessed last year, just over the hills from where I sit now. For me, this would mark the start and finish of the natural winter in these mountains. The virus stopped me wandering back into the hills to find them, yet here they are, only a few feet away. These wild children have come to me.

***

A few days later, back in Inverness, my tent arrives. When I ordered it, I thought it had been manufactured in China, as pretty well everything is. It turned out, however, it actually had to be delivered all the way from China, so it arrives a few days after camping is allowed in Scotland again. The internet is full of horror stories about people flooding into the countryside now that lockdown is eased and leaving trails of destruction behind them. There are stories in the media of crowds in places like Glen Etive near Glen Coe. Some west-coast villages are complaining about the sheer volume of camper vans and finding the sudden influx of people after lockdown unbearable. In the

Cairngorms, the roads around Loch Morlich, near Aviemore, are packed with folk sleeping in their cars. The newspapers publish photographs of the beach clogged with people, and there are fears that campfires will destroy the forest. I desperately want to get back to Sutherland, and hear tales of campsites packed full of people. As bothies are closed, my only option is to wild camp. There is a spot at the foot of Ben Loyal, a few miles from Tongue, that I have often passed and thought what a great campsite it would make. On the drive north, I am struck by how crowded the roads are not. I pass lots of places where the dreaded camper vans could have pulled off the road, but they are all empty. There are fewer people here than in an average summer.

Yet still there are voices of outrage in the local press about visitors wreaking havoc. There is a huge contradiction here. For years, the Highlands have thrown open their doors and tried to cram people in. We have an economy based on tourism, yet now there are angry voices raised about people coming to enjoy the Highland scenery. The media is ever hungry to find villains, to look for people to blame and feed the outrage of its readers. Social media, with its conversations based on brief soundbites, fosters not discussion but polarised arguments. The truth is that we have a population who have been confined to their houses for months. They can hardly be blamed for the fact that now lockdown has eased they want to escape to somewhere green where the air is fresh. Many of these folk are not experienced campers and would normally go abroad for their holidays. It's understandable that they gravitate to the better-known areas and don't understand that campfires are dangerous. Many people's only experience of camping will have been at festivals, where it's normal just to abandon your tent at the end and head home. There is a risk that wild camping will become demonised, and there are even rumours that it will be banned. Mountaineering Scotland has responded to these knee-jerk reactions with a far more reasonable approach, that of educating the public on how to wild-camp in a way that does not wreck the environment. Their Considerate Camping campaign simply aims to tell people about the risks of campfires and how to go to the loo in the wild, and is a good way to educate folk and let us all share the outdoors.

Perhaps we need to rethink our whole approach to bringing people into remote landscapes. When I began camping in the Highlands, Scottish campsites were often not much more than a field with a toilet and a freshwater tap. Over the years, campsites have become grander, with shops and sometimes their own cafes or bars. Many of the lesser campsites have closed, with the result that camping has become more centralised, encouraging visitors to congregate in specific areas and leading to these becoming overcrowded. Marketing initiatives like the North Coast 500 have had a major impact by bringing large numbers of tourists to remote areas without making any investment in the infrastructure to support them. Covid-19 has highlighted the vulnerability of small communities to overpopulation by tourists. Perhaps what we need to do is to encourage a decentralisation and reverse the pressures that have been building up in places like Skye, Glen Coe and the north coast routes. There is plenty of room for everyone as long as we camp responsibly and more local facilities are provided. If we were to set up additional facilities for temporary toilets and rubbish collection points during the high-pressure times of large visitor numbers, while at the same time providing a lot more small campsites with fewer facilities, the strain could be taken off some areas, with the income spread more evenly. Whether this approach would work in places like the Lake District I doubt because these are much smaller areas and the pressure is more intense. I am convinced that there is still room for everyone in the Highlands as long as we look for ways to share this beautiful place.

As soon as I begin to erect my tent, it's immediately obvious to me that I've made a mistake. It's nothing like as sturdy as I thought when I saw it on the internet. It has far fewer guy lines than I imagined. The wind sweeping down the glen is no more than a Sutherland breeze, but even in that my Chinese tent is bucking wildly. I can stand up in the lightweight tepee - and indeed I have to, as I need to hang on to the single central pole to stop it rocking too violently. It's obvious that this tent was designed for high humidity, as it is the most well ventilated tent I have ever seen. The ventilation is allowing the wind in the glen to whistle right through the tent. That this little structure is not up to a Scottish summer is not the manufacturer's fault; I'm sure it was never intended for such conditions. I have no idea what was going through my mind when I bought it.

By the morning, the breeze has grown stronger and it's obvious that I need to get the tent down fast if it is not to end up in Loch Loyal. I'm obviously going to have to buy a much sturdier version if my portable bothy is going to be a viable alternative to the real thing. After I have hurled my tent into the back of my car, I drive along the side of Loch Loyal, heading for a small isthmus between two lochs that am fond of. I leave the car and walk across a small wooden bridge. At last I am in Sutherland again; this is where I saw the otter tracks in November. I lean over the bridge, but today the sand is bare and there are no signs to mark an otter's passage. The wind whips up small waves on the surface of the loch and great grey clouds sweep in over the surrounding hills. I breathe in the smell of the wind and the wild, and I am free at last.

One of my favourite bothies, Achnanclach, is less than two miles away. I could so easily cross the stream with its rickety bridge, take the short walk up the hill and spend the night dreaming as the fire crackles in the hearth. That was another life before all this, I tell myself; I put ideas of the bothy visit out of my mind. I am just lucky to be out here at all. On the far side of this narrow strip of land between the two lochs, there is a sandy beach I've often seen on my walks into the bothy but never visited. As soon as my boots hit the sand, I know I am somewhere special. This is one of those places with an unmistakable magic. There are no footprints on the sand, just the marks of a few birds. The day is windy but the air is warm, and it is comfortable to sit with my back against a bank and watch sand martins skimming across the surface of the bright water. The only sound is the gentle lapping of the small waves against the beach. For this moment, I feel time slow and become part of this whispering, unchanged land. As I sit beside this loch, alone but for the birds and the open sky, the bitter memories of the weeks I spent besieged in my flat slowly drip from my body into the sand. My worries are carried away in the Sutherland wind. The hostile world I saw on the far side of the glass, standing by the window of my flat, has put its arms around me and there is no need to fear. I am home once more.

***

July is almost over and I am walking up into the northern corries of the Cairngorms. They are two bowl-shaped amphitheatres cut deep into the high plateau. Each has a rear wall of steep cliffs with a boulder-strewn floor that slopes gently down the forests of Glenmore. July is a strange time of year to end a book that has charted my passage through winter, but this winter has been stranger than any other. Even though it is summer now, if you know where to look, winter still lingers in these high corries. I climb through the jumble of huge boulders and on to the base of the cliffs. Already I can see what I am looking for lying at the foot of the sheer rock above me. I step through the boulders on to the scree above, and then cross the grass. It is then that my boots crunch into the frozen surface of the winter snow. Here, high in these corries, great patches of snow remain, survivors from the blizzards of March. They rest here like tired old warriors hiding from the summer heat. They are pock-marked with debris from the cliffs above and their faces carry long scars from being struck by fallen rocks. When I press my hand on to the surface of the snow, the icy touch transports me back to November, when James and I lay in the snow watching the sleeping hare. As I walk across the tennis-court-sized patch, my boots crunch on the ice crystals and I am taken back to climbing that hill with Martin and our night beside the fire in Inver hut.

With luck, these patches of snow will lie here until the autumn chill returns and they are replenished by the fleeting snows of November. If our climate were to dip by only one or two degrees, these semi-permanent patches of snow would give birth to the ice kingdoms of rolling glaciers. Even now, streams running beneath them can carve impressive ice caves that harbour an eerie light and are known to the few folk who study these secret bastions of winter.

I pick up a handful of cold stardust and, as it melts and trickles through my fingers, I remember the winter past, the winter we all lived through and none of us will forget. This winter taught me that time melts as quickly as the ice in my hand. I have yet to see a wildcat or watch as a whale breaches in the seas off Mull. I have still to watch a beaver swim by in a Scottish loch or learn the names of the flowers in the hedgerow. These are dreams for seasons to come.

ACKNOWLEDGEMENTS

Thanks to Neil Reid, Alex Roddie, James Roddie, Ben Ross and all at Vertebrate Publishing.

ABOUT THE AUTHOR

John D. Burns is a bestselling and award-winning mountain writer who has spent over forty years exploring Britain's mountains. Originally from Merseyside, he moved to Inverness over thirty years ago to follow his passion for the hills. He is a past member of the Cairngorm Mountain Rescue Team and has walked and climbed in the American and Canadian Rockies, Kenya, the Alps and the Pyrenees.

John began writing more than fifteen years ago, and at first found an outlet for his creativity as a performance poet. He has taken one-man plays to the Edinburgh Fringe and toured them widely around theatres and mountain festivals in the UK.

It is the combination of John's love of the outdoors with his passion for writing and performance that makes him a uniquely powerful storyteller. His first two books, *The Last Hillwalker* and *Bothy Tales*, were both shortlisted for *TGO Magazine*'s Outdoor Book of the Year. His third book, *Sky Dance*, was shortlisted for the Banff Mountain Literature and Poetry award in 2020. He continues to develop his career as a writer, blogger, podcaster and outdoor storyteller while exploring the wild places he loves.

If you would like to follow more of John's journeys to bothies and wild places, and his adventures in the world of theatre, visit *www.johndburns.com* where you can read his blogs and listen to his podcasts.

ALSO BY JOHN D. BURNS

In the twenty-first century, we are losing our connection with the wild, a connection that may never be regained. *The Last Hillwalker* is a personal story of falling in and out of love with the hills. More than that, it is about rediscovering a deeply felt need in all of us to connect with wild places.

In *Bothy Tales*, the follow-up to *The Last Hillwalker*, we travel with the author to remote glens deep in the Scottish Highlands. John brings a second volume of tales – some dramatic, some moving, some hilarious – from the isolated mountain shelters called bothies.

In his first two bestselling books, John invited readers to join him in the hills and wild places of Scotland. In *Sky Dance*, he returns to that world. As wild land is threatened like never before, it's time we asked ourselves what kind of future we want for the Highlands.